Advanced Techniques in
Applied Mathematics

LTCC Advanced Mathematics Series

Series Editors: Shaun Bullett *(Queen Mary University of London, UK)*
Tom Fearn *(University College London, UK)*
Frank Smith *(University College London, UK)*

Published

LTCC Advanced Mathematics Series - Volume 1

Advanced Techniques in
Applied Mathematics

Editors

Shaun Bullett
Queen Mary University of London, UK

Tom Fearn
University College London, UK

Frank Smith
University College London, UK

World Scientific

NEW JERSEY · LONDON · SINGAPORE · BEIJING · SHANGHAI · HONG KONG · TAIPEI · CHENNAI · TOKYO

Published by

World Scientific Publishing Europe Ltd.

57 Shelton Street, Covent Garden, London WC2H 9HE

Head office: 5 Toh Tuck Link, Singapore 596224

USA office: 27 Warren Street, Suite 401-402, Hackensack, NJ 07601

Library of Congress Cataloging-in-Publication Data
Names: Bullett, Shaun, 1967– | Fearn, T., 1949– | Smith, F. T. (Frank T.), 1948–
Title: Advanced techniques in applied mathematics / Shaun Bullett (Queen Mary
    University of London, UK), Tom Fearn (University College London, UK) &
    Frank Smith (University College London, UK).
Description: New Jersey : World Scientific, 2016. | Series: LTCC advanced
    mathematics series ; vol. 1 | Includes bibliographical references and index.
Identifiers: LCCN 2015047092| ISBN 9781786340214 (hc : alk. paper) |
    ISBN 9781786340221 (sc : alk. paper)
Subjects: LCSH: Numerical analysis. | Differential equations. | Differential equations, Partial. |
    Finite element method. | Random matrices.
Classification: LCC QA300 .B83 2016 | DDC 518--dc23
LC record available at http://lccn.loc.gov/2015047092

British Library Cataloguing-in-Publication Data
A catalogue record for this book is available from the British Library.

Desk Editors: Suraj Kumar/Mary Simpson

Typeset by Stallion Press
Email: enquiries@stallionpress.com

Printed in Singapore

Preface

The London Taught Course Centre (LTCC) for PhD students in the Mathematical Sciences has the objective of introducing research students to a broad range of topics. For some students, some of these topics might be of obvious relevance to their PhD projects, but the relevance of most will be much less obvious or apparently non-existent. However all of us involved in mathematical research have experienced that extraordinary moment when the penny drops and some tiny gem of information from outside one's immediate research field turns out to be the key to unravelling a seemingly insoluble problem, or to opening up a new vista of mathematical structure. By offering our students advanced introductions to a range of different areas of mathematics, we hope to open their eyes to new possibilities that they might not otherwise encounter.

Each volume in this series consists of chapters on a group of related themes, based on modules taught at the LTCC by their authors. These modules were already short (five two-hour lectures) and in most cases the lecture notes here are even shorter, covering perhaps three-quarters of the content of the original LTCC course. This brevity was quite deliberate on the part of the editors — we asked the authors to confine themselves to around 35 pages in each chapter, in order to allow as many topics as possible to be included in each volume, while keeping the volumes digestible. The chapters are "advanced introductions", and readers who wish to learn more are encouraged to continue elsewhere. There has been no attempt to make the coverage of topics comprehensive. That would be impossible in any case — any book or series of books which included all that a PhD student in mathematics might need to know would be so large as to be totally unreadable. Instead what we present in this series is a cross-section of some of the topics, both classical and new, that have appeared in LTCC modules in the nine years since it was founded.

The present volume is within the area of advanced techniques in applied mathematics. The main readers are likely to be graduate students and more experienced researchers in the mathematical sciences, looking for introductions to areas with which they are unfamiliar. The mathematics presented is intended to be accessible to first year PhD students, whatever their specialised areas of research. Whatever your mathematical background, we encourage you to dive in, and we hope that you will enjoy the experience of widening your mathematical knowledge by reading these concise introductory accounts written by experts at the forefront of current research.

Shaun Bullett, Tom Fearn, Frank Smith

Contents

Chapter 1

Practical Analytical Methods
for Partial Differential Equations

Helen J. Wilson

Department of Mathematics, University College London,
Gower Street, London WC1E 6BT, UK
helen.wilson@ucl.ac.uk

This chapter runs through some techniques that can be used to tackle partial differential equations (PDEs) in practice. It is not a theoretical work — there will be no proofs — instead I will demonstrate a range of tools that you might want to try. We begin with first-order PDEs and the method of characteristics; classification of second-order PDEs and solution of the wave equation; and separation of variables. Finally, there is a section on perturbation methods which can be applicable to both ordinary differential equations (ODEs) and PDEs of any order as long as there is a small parameter.

1. Introduction

We will see a variety of techniques for solving, or approximating the solution of, differential equations. Each is illustrated by means of a simple example. In many cases, these examples are so simple that they could have been solved by simpler methods; but it is instructive to see new methods applied without having to wrestle with technical difficulties at the same time.

Section 2 deals with first-order equations. In Section 3, we classify second-order partial differential equations (PDEs) into hyperbolic, parabolic, and elliptic; then hyperbolic equations are tackled in Section 4 and we briefly discuss elliptic equations in Section 5. Section 6 reviews the well-known theory of separation of variables. Finally, in Section 7 we develop the theory of matched asymptotic expansions, suitable for use in PDEs having a small parameter.

The principal text for most of the chapter is by Weinberger [1]; though the book is out of print the full text is freely available online. In the later material on asymptotic expansions, there are several relevant texts, including those by Bender and Orszag [2], Kevorkian and Cole [3], and Van Dyke [4]. My presentation is most similar to that by Hinch [5].

2.   First-order PDEs

First-order partial differential equations can be tackled with the **method of characteristics**. We will develop the method from the simplest case first: a constant-coefficient linear equation.

2.1.   *Wave equation with constant speed*

The first-order wave equation with constant speed:

$$\partial u/\partial t + c\,\partial u/\partial x = 0$$

responds well to a change of variables:

$$\xi = x + ct; \quad \eta = x - ct.$$

The extended chain rule gives us

$$\frac{\partial}{\partial x} = \frac{\partial \xi}{\partial x}\frac{\partial}{\partial \xi} + \frac{\partial \eta}{\partial x}\frac{\partial}{\partial \eta} = \frac{\partial}{\partial \xi} + \frac{\partial}{\partial \eta};$$

$$\frac{\partial}{\partial t} = \frac{\partial \xi}{\partial t}\frac{\partial}{\partial \xi} + \frac{\partial \eta}{\partial t}\frac{\partial}{\partial \eta} = c\left(\frac{\partial}{\partial \xi} - \frac{\partial}{\partial \eta}\right)$$

and so the wave equation is equivalent to

$$2c\,\partial u/\partial \xi = 0.$$

Integrating gives the general solution $u = F(\eta)$, $u = F(x - ct)$.

2.2.   *Characteristics*

Where did we get the change of variables from? We can see that, in our choice of variables, only the definition of η is important. Any ξ (independent of η) would be fine as the other variable; since u is a function of η, differentiating while holding η constant will always give zero.

Thus the important thing about our change of variables is the family of lines defined by $\eta =$ constant. These lines are called **characteristics** and they are a property of the differential operator on the left of our equation. For a homogeneous equation like this, they are lines along which the solution is constant; but they are critically important for all first-order PDEs.

To find the characteristics for a given first-order differential operator, we look for a change of variables under which the whole operator transforms to a multiple of single derivative. If we have the operator

$$\mathcal{L}[u] = \frac{\partial u}{\partial t} + c(x,t)\frac{\partial u}{\partial x},$$

one way to do this is to make the operator mimic the chain rule:

$$\frac{\partial}{\partial r} = \frac{\partial t}{\partial r}\frac{\partial}{\partial t} + \frac{\partial x}{\partial r}\frac{\partial}{\partial x}, \quad \frac{\partial t}{\partial r} = 1, \quad \frac{\partial x}{\partial r} = c(x,t).$$

We should be able to integrate both of these in turn. The variable r parametrizes each of our characteristic curves; and there is **one** constant of integration, x_0 say, which serves to label the curves.

Example

$$2\sin^2\theta\cos 2\phi\frac{\partial u}{\partial\theta} - \cos\theta\sin 2\phi\frac{\partial u}{\partial\phi} = \cos 2\phi(u\cos\theta + \cot\theta).$$

The operator here is

$$\mathcal{L}[u] = 2\sin^2\theta\cos 2\phi\frac{\partial u}{\partial\theta} - \cos\theta\sin 2\phi\frac{\partial u}{\partial\phi}.$$

Here, it makes sense to uncouple the equations we will have to solve to find the characteristics, by first dividing the original operator by $\cos\theta\cos 2\phi$:

$$\frac{1}{\cos\theta\cos 2\phi}\mathcal{L}[u] = \frac{2\sin^2\theta}{\cos\theta}\frac{\partial u}{\partial\theta} - \frac{\sin 2\phi}{\cos 2\phi}\frac{\partial u}{\partial\phi},$$

$$\frac{\partial\theta}{\partial r} = \frac{2\sin^2\theta}{\cos\theta}, \quad \frac{\partial\phi}{\partial r} = -\frac{\sin 2\phi}{\cos 2\phi}.$$

Solving the equations in turn gives

$$\sin\theta = -1/(2r), \quad \sin 2\phi = \exp[x_0 - 2r].$$

We only need to use a constant of integration in one of these equations: since r is just a parameter, the point $r = 0$ is not defined *a priori*. We have now found our change of variables from $\{\theta, \phi\}$ to $\{r, x_0\}$. We can invert the transformation:

$$x_0 = \ln(\sin 2\phi) - 1/\sin\theta, \quad r = -1/(2\sin\theta)$$

and the operator becomes

$$\mathcal{L}[u] = \cos\theta \cos 2\phi \left.\frac{\partial u}{\partial r}\right|_{\text{constant } x_0}.$$

Returning to the whole PDE, we now have

$$\cos\theta \cos 2\phi \frac{\partial u}{\partial r} = \cos 2\phi \left(u\cos\theta + \cot\theta\right); \quad \frac{\partial u}{\partial r} = u - 2r,$$

$$u = F(x_0)e^r + 2r + 2 = F\left(\ln\sin 2\phi - \frac{1}{\sin\theta}\right)\exp\left[\frac{-1}{2\sin\theta}\right] + 2 - \frac{1}{\sin\theta}.$$

2.3. *Nonlinear equations*

We are now looking at the most general first-order PDEs: those of the form

$$\frac{\partial u}{\partial t} + c(u, x, t)\frac{\partial u}{\partial x} = f(u, x, t).$$

Characteristics still exist in these systems, and they have important physical properties (for instance, discontinuities in the derivatives of the solution will propagate along them) but we have no explicit solution method.

However, if the equation is homogeneous so that $f(u, x, t) = 0$, we can solve implicitly. First assume that we have found the characteristics $t = r$, $x(x_0, r)$ satisfying

$$\partial x/\partial r = c(u, x, r).$$

Then along any characteristic we will have $\partial u/\partial r = 0$ so that $u = F(x_0)$. Now return to the equation governing the characteristic:

$$\partial x/\partial r = c(F(x_0), x, r);$$

this can now be treated as a straightforward ordinary differential equation (ODE) in x and r. Once we have solved it we have the characteristic curve

$$t = r \quad x = G(x_0, F(x_0), r) = G(x_0, u, r).$$

The implicit form of the solution is now:

$$x = G(x_0, u, t), \quad u = F(x_0).$$

Example

Consider the advection equation

$$\partial u/\partial t + ux^2t\,\partial u/\partial x = 0.$$

As u is constant on each characteristic, we set $u = F(x_0)$ and the equation governing our characteristic curve becomes

$$\partial x/\partial r = ux^2r = F(x_0)x^2r,$$

which we can solve:

$$\int \frac{\mathrm{d}x}{x^2} = F(x_0) \int r\,\mathrm{d}r, \quad x = \frac{2x_0}{2 - x_0F(x_0)r^2}.$$

Thus, the characteristic curve and implicit solution are:

$$t = r, \quad x = \frac{2x_0}{2 - x_0F(x_0)r^2}, \quad u = F(x_0).$$

In this case we can rearrange to get x_0 in terms of x, t, and u, which gives the standard implicit form of the solution:

$$u = F\left(\frac{2x}{2 + uxt^2}\right).$$

3. Linear second-order PDEs: Classification

Consider a linear homogeneous second-order PDE in x and t having constant coefficients and only second-order derivatives:

$$A\frac{\partial^2 f}{\partial t^2} + B\frac{\partial^2 f}{\partial x\partial t} + C\frac{\partial^2 f}{\partial x^2} = 0; \quad \mathcal{L}[f] := A\frac{\partial^2 f}{\partial t^2} + B\frac{\partial^2 f}{\partial x\partial t} + C\frac{\partial^2 f}{\partial x^2}.$$

We have introduced the linear differential operator $\mathcal{L}$, and we can reduce it to one of three canonical forms using a linear coordinate transformation: if we take

$$\xi = \alpha x + \beta t, \quad \eta = \gamma x + \delta t,$$

the operator becomes

$$\mathcal{L}[f] = [A\beta^2 + B\alpha\beta + C\alpha^2]\frac{\partial^2 f}{\partial \xi^2}$$

$$+ [2A\beta\delta + B(\alpha\delta + \beta\gamma) + 2C\alpha\gamma]\frac{\partial^2 f}{\partial \xi \partial \eta} + [A\delta^2 + B\gamma\delta + C\gamma^2]\frac{\partial^2 f}{\partial \eta^2}.$$

As we will see in Section 4, it can be very helpful to choose coordinates in which only the second term has nonzero coefficient. Can we manage this in general? We will need to choose α, β, γ, and δ so that

$$A\beta^2 + B\alpha\beta + C\alpha^2 = 0, \quad A\delta^2 + B\gamma\delta + C\gamma^2.$$

Suppose for the sake of the argument that A is nonzero. Then neither α nor γ can be zero and we can divide by them:

$$A\left(\frac{\beta}{\alpha}\right)^2 + B\left(\frac{\beta}{\alpha}\right) + C = 0, \quad A\left(\frac{\delta}{\gamma}\right)^2 + B\left(\frac{\delta}{\gamma}\right) + C = 0.$$

The solutions (identical) to these two constraints are

$$\frac{\beta}{\alpha} = \frac{1}{2A}[-B \pm \sqrt{B^2 - 4AC}], \quad \frac{\delta}{\gamma} = \frac{1}{2A}[-B \pm \sqrt{B^2 - 4AC}]$$

and the two ratios must be different (for otherwise ξ is a multiple of η): so providing $B^2 - 4AC > 0$ we can set $\alpha = \gamma = 2A$ and choose

$$\xi = 2Ax + [-B + \sqrt{B^2 - 4AC}]t, \quad \eta = 2Ax + [-B - \sqrt{B^2 - 4AC}]t,$$

which reduces the whole equation to

$$\mathcal{L}[f] = -4A(B^2 - 4AC)\frac{\partial^2 f}{\partial\xi\partial\eta} = 0,$$

which is easy to solve.

This only works where $B^2 - 4AC > 0$: and in fact this quantity, the **discriminant**, is very powerful in determining the global behaviour of the PDE solution. There are three possible cases.

Discriminant positive: Hyperbolic system

This is the case we have just looked at: it can be reduced to a single mixed-derivative term. In Section 4, we will look in detail at the canonical example of the second-order wave equation

$$\frac{\partial^2 u}{\partial t^2} - c^2\frac{\partial^2 u}{\partial x^2} = 0,$$

which has $A = 1$, $B = 0$, and $C = -c^2$ giving a discriminant of $4c^2$. These systems have two families of characteristics along which information propagates, and a typical solution is a combination of travelling waves.

Discriminant zero: Parabolic system

If the discriminant is zero, $B^2 = 4AC$, we can use the new variables $\xi = 2Ax - Bt$ and any η independent of ξ, and the PDE becomes

$$[A\delta^2 + B\gamma\delta + C\gamma^2]\frac{\partial^2 f}{\partial \eta^2} = 0.$$

A typical parabolic system is the steady 1D heat equation

$$\partial^2 u/\partial x^2 = 0;$$

but another parabolic equation, which is of immense importance in applications, is the time-dependent heat equation

$$\partial u/\partial t = \kappa \partial^2 u/\partial x^2.$$

Discriminant negative: Elliptic system

In this case we cannot get rid of the coefficients of $u_{\xi\xi}$ and $u_{\eta\eta}$; however, we can eliminate the mixed derivative using

$$\xi = 2Ax - Bt, \qquad \eta = \sqrt{4AC - B^2}t,$$

to obtain

$$A(4AC - B^2)\left[\frac{\partial^2 f}{\partial \xi^2} + \frac{\partial^2 f}{\partial \eta^2}\right] = 0.$$

The classic example of an elliptic PDE is precisely this reduced form: Laplace's equation

$$\frac{\partial^2 u}{\partial t^2} + \frac{\partial^2 u}{\partial x^2} = 0, \quad \frac{\partial^2 u}{\partial x^2} + \frac{\partial^2 u}{\partial y^2} = 0, \quad \nabla^2 u = 0,$$

whose discriminant is -1. There are no real characteristics in this case, though complex characteristics are a powerful abstract tool. Typical solutions are energy-minimizing surfaces or functions.

3.1. *Varying coefficients*

If we have a general linear second-order partial differential operator:

$$\mathcal{L}[f] = A(x,t)\frac{\partial^2 f}{\partial t^2} + B(x,t)\frac{\partial^2 f}{\partial x \partial t} + C(x,t)\frac{\partial^2 f}{\partial x^2}$$

$$+ D(x,t)\frac{\partial f}{\partial t} + E(x,t)\frac{\partial f}{\partial x} + F(x,t)f,$$

then it is hyperbolic, parabolic or elliptic *at a point* (x_0, t_0) according to the value of $B^2 - 4AC$ at that point, in other words the local discriminant.

4.   Hyperbolic PDE: The second-order 1D wave equation

4.1.   *Homogeneous wave equation with constant speed*

This wave equation is the simplest hyperbolic partial differential equation:

$$\frac{\partial^2 u}{\partial t^2} - c^2 \frac{\partial^2 u}{\partial x^2} = 0.$$

Like the first-order wave equation, it responds well to a change of variables:

$$\xi = x + ct, \quad \eta = x - ct.$$

This reduces it to

$$-4c^2 \frac{\partial^2 u}{\partial \xi \partial \eta} = 0 \quad \Rightarrow \quad u = p(\xi) + q(\eta) = p(x + ct) + q(x - ct)$$

for any differentiable functions p and q. The lines of constant ξ or η are the characteristics, exactly analogous to those for the first-order equation.

If we add initial conditions $u(x, 0) = f(x)$ and $\partial u / \partial t(x, 0) = g(x)$ then a little algebra gives us **d'Alembert's solution**:

$$u(x, t) = \frac{1}{2}\left[f(x + ct) + f(x - ct)\right] + \frac{1}{2c} \int_{x-ct}^{x+ct} g(y)\, dy.$$

4.2.   *Inhomogeneous wave equation*

The inhomogeneous wave equation:

$$\frac{\partial^2 u}{\partial t^2} - c^2 \frac{\partial^2 u}{\partial x^2} = F(x, t)$$

can be solved in a very similar way. The change of variables results in:

$$-4c^2 \frac{\partial^2 u}{\partial \xi \partial \eta} = F\left(\frac{\xi + \eta}{2}, \frac{\xi - \eta}{2c}\right),$$

which can be integrated directly for any specific function F; however (Weinberger [1], p. 25) it is also possible to carry out the integrals symbolically. The general result is

$$u = p(x + ct) + q(x - ct) + \frac{1}{2c} \int_0^t \int_{x-c(t-t')}^{x+c(t-t')} F(x', t')\, dx'\, dt'. \quad (1)$$

Example

Let us consider the example inhomogeneous wave equation

$$\frac{\partial^2 u}{\partial t^2} - c^2 \frac{\partial^2 u}{\partial x^2} = 12xt.$$

The easiest way to solve is to make the change of variables directly: when we put $\xi = x + ct$ and $\eta = x - ct$ we obtain

$$-4c^2 \frac{\partial^2 u}{\partial \xi \partial \eta} = \frac{3}{c}(\xi + \eta)(\xi - \eta) = \frac{3}{c}(\xi^2 - \eta^2).$$

Integrating twice (once with respect to each variable) gives

$$-4c^2 u = p(\xi) + q(\eta) + \frac{\xi \eta}{c}(\xi + \eta)(\xi - \eta)$$

and converting the coordinates back gives the full solution:

$$u = f(x + ct) + g(x - ct) - \frac{xt(x^2 - c^2 t^2)}{c^2}.$$

Exercise for the reader: solve using Equation (1) and show that the difference between the two solutions can be absorbed into f and g above.

4.3.   *Varying speed: Two sets of characteristics*

We saw in the constant-speed case that the characteristic curves were the straight lines

$$\xi = x + ct = \text{constant}, \quad \eta = x - ct = \text{constant}.$$

Any point (x, t) lies on two characteristics; even for a homogeneous equation the function is not constant along the characteristics. Instead, information propagates along characteristics.

Figure 1 shows the characteristics passing through the point $(\bar{x}, \bar{t})$ for a general homogeneous hyperbolic equation. These characteristics C_1 and C_2 meet the line $t = 0$ at x_1 and x_2 respectively. Because information only propagates along characteristics, the value of the solution at $(\bar{x}, \bar{t})$ depends only on the initial conditions between x_1 and x_2 and not elsewhere.

Weinberger [1] proves this (pp. 37–38) for the specific case of the wave equation whose wavespeed varies in space:

$$\frac{\partial^2 u}{\partial t^2} - c^2(x) \frac{\partial^2 u}{\partial x^2} = 0.$$

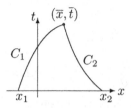

Fig. 1.   Characteristics leading to the point $(\bar{x}, \bar{t})$.

4.4.   *General hyperbolic equation*

Suppose that the discriminant is positive everywhere in our domain, then we say the PDE is **hyperbolic in the domain,** and it will have two families of characteristics along which information propagates. We can reduce the second-order terms to the standard form $\partial^2 u/\partial \xi \partial \eta$, although that does not guarantee us a solution: nonetheless, finding the characteristic curves can also be very useful for numerical solution.

Instead of using a linear change of variables, we use a general (twice differentiable) transformation:

$$\xi = \xi(x,t), \quad \eta = \eta(x,t)$$

and the chain rule gives:

$$\mathcal{L}[f] = \left[A\left(\frac{\partial \xi}{\partial t}\right)^2 + B\frac{\partial \xi}{\partial t}\frac{\partial \xi}{\partial x} + C\left(\frac{\partial \xi}{\partial x}\right)^2 \right]\frac{\partial^2 f}{\partial \xi^2}$$

$$+ \left[2A\frac{\partial \xi}{\partial t}\frac{\partial \eta}{\partial t} + B\frac{\partial \xi}{\partial t}\frac{\partial \eta}{\partial x} + B\frac{\partial \eta}{\partial t}\frac{\partial \xi}{\partial x} + 2C\frac{\partial \xi}{\partial x}\frac{\partial \eta}{\partial x} \right]\frac{\partial^2 f}{\partial \xi \partial \eta}$$

$$+ \left[A\left(\frac{\partial \eta}{\partial t}\right)^2 + B\frac{\partial \eta}{\partial t}\frac{\partial \eta}{\partial x} + C\left(\frac{\partial \eta}{\partial x}\right)^2 \right]\frac{\partial^2 f}{\partial \eta^2}$$

$$+ \text{terms in } \frac{\partial f}{\partial \xi}, \frac{\partial f}{\partial \eta}, f.$$

To zero the unmixed second derivatives, both ratios

$$\frac{\partial \xi/\partial t}{\partial \xi/\partial x}, \quad \frac{\partial \eta/\partial t}{\partial \eta/\partial x}$$

need to be solutions of the quadratic equation $Az^2 + Bz + C = 0$; we choose one root for each family of curves and solve from there.

A word of warning here: the inverse transform from the characteristic variables back to the original variables can become singular; this happens, for example, when describing breaking waves.

5. Elliptic second-order PDEs: Laplace's equation

The most common elliptic second-order PDE is Laplace's equation:

$$\frac{\partial^2 f}{\partial x^2} + \frac{\partial^2 f}{\partial y^2} = 0.$$

It arises in a huge variety of physical scenarios, from fluid dynamics to electromagnetism and astronomy. And, indeed, with a suitable change of variables any homogeneous elliptic second-order PDE with constant coefficients can be transformed into Laplace's equation.

It has no real characteristics because its discriminant is negative ($B^2 - 4AC = -4$). But if we ignore this technicality and allow ourselves a complex change of variables, we can benefit from the same structure of solution that worked for the wave equation. Introduce

$$z = x + iy, \quad x = (z + \overline{z})/2,$$
$$\overline{z} = x - iy, \quad y = (z - \overline{z})/2i.$$

Then the PDE becomes

$$4\frac{\partial}{\partial \overline{z}}\frac{\partial f}{\partial z} = 0.$$

whose solution is straightforward: $f = p(z) + q(\overline{z})$. Here, p and q are differentiable complex functions; and assuming we wanted a real solution to the original (real) PDE, we have an additional constraint that the sum of the two functions must have no imaginary part.

It is then possible to show that f must, in fact, be the real part of an analytic function of z, and this leads us to the elegant theory of complex variables and conformal maps. This is described in, amongst others, two excellent textbooks by Brown and Churchill [6] and Nehari [7].

6. General linear PDEs: Separation of variables, a "Lucky" method

Separation of variables is hugely useful when it works, but can only work in rather special cases. It depends critically on the linearity of your problem and the type of your boundary conditions; if these are favourable then it is always worth a try as it is pretty easy to use.

6.1. The basics

We seek to express our solution as a sum of solutions of the form

$$f(x, t) = X(x)T(t).$$

We substitute this into the governing equation, and divide by f. If the coefficients behave well, each term will be either a function of x or of t. Then we move all functions of x to one side of the equation, and functions of t to the other:

$$F(x) = G(t).$$

Since this is true for all x and t, it follows that each side of the equation must be constant. We name the constant (say A) and then have two ODEs to solve:

$$F(x) = A, \quad G(t) = A.$$

Then these pairs of solutions (different for different values of A) can be superposed to give the general solution.

6.1.1. Example: Laplace in plane polars

Laplace's equation in plane polar coordinates is

$$\frac{1}{r}\frac{\partial}{\partial r}\left(r\frac{\partial f}{\partial r}\right) + \frac{1}{r^2}\frac{\partial^2 f}{\partial \theta^2} = 0, \quad r^2\frac{\partial^2 f}{\partial r^2} + r\frac{\partial f}{\partial r} + \frac{\partial^2 f}{\partial \theta^2} = 0.$$

The separable solution $f(r, \theta) = R(r)T(\theta)$ gives the coupled ODEs

$$\frac{r^2 R''(r) + rR'(r)}{R(r)} = A, \quad \frac{T''(\theta)}{T(\theta)} = -A,$$

$$r^2 R''(r) + rR'(r) - AR(r) = 0, \quad T''(\theta) = -AT(\theta).$$

We look at the three cases $A > 0$, $A < 0$ and $A = 0$ separately.

Positive constant $A = \lambda^2$
$$R(r) = a_1 r^\lambda + a_2 r^{-\lambda}, \ T(\theta) = a_3 \cos \lambda\theta + a_4 \sin \lambda\theta.$$

Zero constant $A = 0$
$$R(r) = b_1 + b_2 \ln r, \ T(\theta) = b_3 + b_4 \theta.$$
Negative constant $A = -\mu^2$
$$R(r) = c_1 \cos(\mu \ln r) + c_2 \sin(\mu \ln r), \ T(\theta) = c_3 e^{\mu\theta} + c_4 e^{-\mu\theta}.$$

Suppose our domain encircles the origin: each function of θ must then be 2π-periodic. This fixes λ to be an integer and b_4 and all the c_i to be zero. The general solution to Laplace's equation in this domain is then:

$$f(r,\theta) = A + B \ln r + \sum_n (C_n r^n + D_n r^{-n})(E_n \cos n\theta + F_n \sin n\theta).$$

6.2. *Boundary conditions*

Of course, Laplace's equation is also separable in Cartesian coordinates; so how do we know which coordinates to use?

The simple answer is that the boundary conditions are crucial. We need the following conditions to be satisfied:

Separable equation
  The differential equation must be separable: that is, it is linear, there are no mixed derivatives[a] and, if the coefficients depend on η and ξ, then (after multiplication of the whole equation by some function if necessary) the derivatives w.r.t. η have coefficients which depend only on η and those w.r.t. ξ have coefficients which depend only on ξ. The coefficient of the no-derivatives term must be at worst the sum of a function of η and a function of ξ.

Boundary conditions on coordinate lines
  All the boundary conditions in the problem must be located along lines $\eta = $ constant or $\xi = $ constant. This does include the possibility of a boundary condition as one variable $\to \infty$.

Correct type of boundary conditions
  Along a line $\eta = $ constant, the boundary condition must not involve any partial derivatives with respect to ξ; and the coefficients of derivatives involved in the boundary conditions must not vary with ξ. The equivalent condition is required of the boundary conditions along a line $\xi = $ constant.

Realistically, the boundary conditions are likely to completely constrain the coordinates we use if we wish to use separation of variables; and if

---

[a]This condition may be relaxed; see p. 70 of Weinberger [1].

the coordinates that work for the boundary conditions do not work for the PDE, there is very little we can do about it.

7. Systems having a small parameter: Perturbation methods

In this section we will look at matched asymptotic expansions, a technique arising from the study of high-Reynolds number fluid mechanics, but which is widely applicable wherever there is a small parameter. Throughout this section, $0 < \varepsilon \ll 1$.

7.1. *Regular perturbation expansions*

We are all familiar with the principle of the Taylor expansion: for an analytic function $f(x)$, we can expand close to a point $x = a$ as:

$$f(a + \varepsilon) = f(a) + \varepsilon f'(a) + \tfrac{1}{2}\varepsilon^2 f''(a) + \cdots.$$

This behaviour, in which a small change to x makes a small change to $f(x)$, is the basis of regular perturbation expansions.

To carry out a regular perturbation expansion is straightforward:

(1) Set $\varepsilon = 0$ and solve the resulting system (name the solution f_0).
(2) Perturb the system by allowing ε to be nonzero (but small).
(3) Formulate the solution to the new, perturbed system as a series

$$f_0 + \varepsilon f_1 + \varepsilon^2 f_2 + \cdots.$$

(4) Expand the governing equations as a series in ε, collecting terms with equal powers of ε; solve them in turn as far as the solution is required.

Example

Suppose we are trying to solve the following differential equation in $x \geq 0$:

$$\mathrm{d}f(x)/\mathrm{d}x + f(x) - \varepsilon f^2(x) = 0, \quad f(0) = 2. \tag{2}$$

We look first at $\varepsilon = 0$:

$$f'(x) + f(x) = 0, \quad f(0) = 2, \quad \Rightarrow \quad f_0(x) = 2e^{-x}.$$

Now we follow our system and set

$$f = f_0(x) + \varepsilon f_1(x) + \varepsilon^2 f_2(x) + \varepsilon^3 f_3(x) + \cdots.$$

where in order to satisfy the initial condition $f(0) = 2$, we will have $f_1(0) = f_2(0) = f_3(0) = \cdots = 0$. Substituting into Equation (2) gives

$$
\begin{aligned}
-2e^{-x} + \varepsilon f_1'(x) + & \quad \varepsilon^2 f_2'(x) \\
+2e^{-x} + \varepsilon f_1(x) + & \quad \varepsilon^2 f_2(x) \\
- 4\varepsilon e^{-2x} - 4\varepsilon^2 e^{-x} f_1(x) &= O(\varepsilon^3)
\end{aligned}
$$

and we can collect powers of ε:

$$
\begin{aligned}
\varepsilon^0 : & \quad -2e^{-x} + 2e^{-x} = 0, \\
\varepsilon^1 : & \quad f_1'(x) + f_1(x) - 4e^{-2x} = 0, \\
\varepsilon^2 : & \quad f_2'(x) + f_2(x) - 4e^{-x} f_1(x) = 0.
\end{aligned}
$$

The order ε^0 (or 1) equation is satisfied automatically. Now we simply solve at each order, applying the boundary conditions as we go along.

Order ε: $f_1'(x) + f_1(x) = 4e^{-2x} \Rightarrow f_1(x) = 4(e^{-x} - e^{-2x})$.
Order ε^2: $f_2'(x) + f_2(x) = 16e^{-x}(e^{-x} - e^{-2x})$
$$\Rightarrow f_2(x) = 8(e^{-x} - 2e^{-2x} + e^{-3x}).$$

The solution we have found is:

$$
f(x) = 2e^{-x} + 4\varepsilon(e^{-x} - e^{-2x}) + 8\varepsilon^2(e^{-x} - 2e^{-2x} + e^{-3x}) + \cdots.
$$

7.1.1. *Warning signs*

This is pleasingly simple, but real life is not always straightforward. A regular perturbation series is often not enough to capture a system's behaviour. Here, are a few of the possible warning signs that things might be going wrong:

One of the powers of ε produces an insoluble equation
By this I do not mean a differential equation with no analytic solution: that is just bad luck. Rather I mean an equation of the form $x_1 + 1 - x_1 = 0$ which cannot be satisfied by any value of x_1. This is a symptom of having the wrong expansion series; see Section 7.3.

The equation at $\varepsilon = 0$ does not give the right number of solutions
An nth order ODE should have n solutions. If the equation produced by setting $\varepsilon = 0$ has fewer solutions, then this method will not give all the possible solutions to the full equation. This happens when the

coefficient of the highest derivative is zero when $\varepsilon = 0$. Equally, for a PDE, if the coefficient of the highest derivative with respect to *one of* the variables is zero when $\varepsilon = 0$, you will not be able to satisfy all your boundary conditions and a regular expansion will not be enough. This is a sign of needing to change the scalings at leading-order; see Section 7.2 and the material following on from it.

The coefficients of ε can grow without bound

In the case of an expansion $f(x) = f_0(x) + \varepsilon f_1(x) + \varepsilon^2 f_2(x) + \cdots$, the series may not be valid for some values of x if some or all of the $f_i(x)$ become very large. Say, for example, that $f_2(x) \to \infty$ while $f_1(x)$ remains finite. Then $\varepsilon f_1(x)$ is no longer strictly larger than $\varepsilon^2 f_2(x)$ and who knows what even larger terms we may have neglected? This is a sign that we need to stretch the underlying variable: see Section 7.4.

7.2. *Not enough solutions: Singular expansion*

In this section, we will look at one of the reasons that our $\varepsilon = 0$ system might not have enough solutions, and introduce a tool that is fundamental to all perturbation systems. We will start with a very simple example and work up from there.

Example

Here, our model equation is

$$\varepsilon x^2 + x - 1 = 0. \tag{3}$$

Suppose we try a regular perturbation expansion on it. Setting $\varepsilon = 0$ gives

$$x - 1 = 0,$$

with just the one solution $x = 1$. Since we started with a second-degree polynomial we know we have lost one of our solutions; however, if we carry on with the regular perturbation expansion we will get a perfectly valid series for the root near $x = 1$.

Now let us look at the true solution to see what has gone wrong.

$$x = \frac{-1 \pm \sqrt{1 + 4\varepsilon}}{2\varepsilon}.$$

As $\varepsilon \to 0$, the leading-order terms of the two roots are

$$x = 1 + O(\varepsilon); \quad \text{and} \quad -\frac{1}{\varepsilon} + O(1).$$

The first of these is amenable to the simplistic approach; we have not seen the second root because it tends to infinity as $\varepsilon \to 0$.

For this second root, let us try a series

$$x = x_{-1}\varepsilon^{-1} + x_0 + \varepsilon x_1 + \cdots .$$

We substitute it into Equation (3):

$$
\begin{aligned}
x_{-1}^2\varepsilon^{-1} + 2x_{-1}x_0 + \varepsilon(x_0^2 + 2x_{-1}x_1) + \cdots & \\
+ x_{-1}\varepsilon^{-1} + x_0 + \varepsilon x_1 + \cdots & \\
- 1 & = 0
\end{aligned}
$$

and collecting powers of ε gives:

$$
\begin{aligned}
\varepsilon^{-1} : \quad & x_{-1}^2 + x_{-1} = 0; \ x_{-1} = \quad 0, -1. \\
\varepsilon^0 : \quad & 2x_{-1}x_0 + x_0 - 1 = 0; \quad x_0 = \quad 1, -1. \\
\varepsilon^1 : \quad & x_0^2 + 2x_{-1}x_1 + x_1 = 0; \quad x_1 = -1, \quad 1.
\end{aligned}
$$

We now get the expansions for both of the roots using the same method.

7.2.1. *Finding the scaling*

What do we do if we cannot use the exact solution to tell us about the first term in the series? We use a trial scaling δ. We put

$$x = \delta(\varepsilon)X$$

with δ being an unknown function of ε, and X being strictly order 1. We call this $X = \text{ord}(1)$: as $\varepsilon \to 0$, X is neither small nor large.

Let us try it for our example equation: $\varepsilon x^2 + x - 1 = 0$. We put in the new form:

$$\varepsilon\delta^2 X^2 + \delta X - 1 = 0$$

and then look at the different possible values of δ. We will only get an order 1 solution for X if the biggest term in the equation is the same size as another term: a **dominant balance** or **distinguished scaling**.

Finding scalings in large systems is more of an art than a science — it is easy to check your scaling works, but finding it in the first place is tricky. But with small systems, it is quite straightforward. I view this process in two ways: one completely systematic (but really only practical with a three-term equation) and the other more of a mental picture.

Systematic method

Since we need the two largest terms to balance, we try all the possible pairs of terms and find the value of δ at which they are the same size. Then for each pair we check that the other term is not bigger than our balancing size.

Balance terms 1 and 2: These two are the same size when $\varepsilon\delta^2 = \delta$ which gives $\delta = \varepsilon^{-1}$. Then both terms 1 and 2 scale as ε^{-1} and term 3 is smaller — so this scaling works.

Balance terms 1 and 3: These two balance when $\varepsilon\delta^2 = 1$ and so $\delta = \varepsilon^{-1/2}$. Then our two terms are both order 1, and term 2 scales as $\varepsilon^{-1/2}$ which is bigger. A single term dominates; this is no good.

Balance terms 2 and 3: These two balance when $\delta = 1$, when they are both order 1. Then term 1 is order ε, which is smaller: so we have a working balance at $\delta = 1$.

The only two scalings which work are $\delta = \varepsilon^{-1}$ and $\delta = 1$.

Horse-race picture

Think of the terms as horses, which "race" as we change δ. The largest term is considered to be leading, and we are interested in the moment when the lead horse is overtaken: that is, the two biggest terms are equal in size.

The three horses in our case are

$$[\mathbf{A}] \quad \varepsilon\delta^2 \quad [\mathbf{B}] \quad \delta \quad [\mathbf{C}] \quad 1$$

and we will start from the point $\delta \approx 0$. Initially, $[\mathbf{C}]$ is ahead, with $[\mathbf{B}]$ second and $[\mathbf{A}]$ a distant third.

As we increase δ, each horse moves according to its power of δ: higher powers move faster (but start further behind). We are looking for the first moment that one of $[\mathbf{A}]$ or $[\mathbf{B}]$ catches $[\mathbf{C}]$. A quick glance tells us that for $[\mathbf{B}]$ it will happen at $\delta = 1$ whereas for $[\mathbf{A}]$ we have to wait until $\delta > 1$. So the first balance is at $\delta = 1$, when $[\mathbf{B}]$ overtakes $[\mathbf{C}]$.

Now because $[\mathbf{C}]$ is the slowest horse (in fact stationary) it will never catch $[\mathbf{B}]$ again, so we only need to look for the moment (if any) when $[\mathbf{A}]$ overtakes $[\mathbf{B}]$. This is given by $\varepsilon\delta^2 = \delta$ which gives our second balancing scaling of $\delta = \varepsilon^{-1}$.

7.3. *Insoluble equations: Different expansion series*

Try this algebraic equation:

$$(1 - \varepsilon)x^2 - 2x + 1 = 0.$$

Setting $\varepsilon = 0$ gives a double root $x = 1$. Now we try an expansion:

$$x = 1 + \varepsilon x_1 + \varepsilon^2 x_2 + \cdots .$$

Substituting in gives

$$
\begin{aligned}
1 + 2\varepsilon x_1 &+ \varepsilon^2(x_1^2 + 2x_2) + \cdots \\
- \varepsilon &- 2\varepsilon^2 x_1 \quad + \cdots \\
- 2 - 2\varepsilon x_1 &- 2\varepsilon^2 x_2 \quad + \cdots \\
+ 1 & \quad\quad\quad = 0
\end{aligned}
$$

At ε^0, as expected, the equation is automatically satisfied. However, at order ε^1, the equation is

$$2x_1 - 1 - 2x_1 = 0, \quad 1 = 0.$$

which we can never satisfy. Something has gone wrong.

In fact in this case we should have expanded in powers of $\varepsilon^{1/2}$. If we set

$$x = 1 + \varepsilon^{1/2} x_{1/2} + \varepsilon x_1 + \cdots$$

then we get

$$
\begin{aligned}
1 + 2\varepsilon^{1/2} x_{1/2} &+ \varepsilon(x_{1/2}^2 + 2x_1) + \cdots \\
&- \varepsilon \quad\quad + \cdots \\
- 2 - 2\varepsilon^{1/2} x_{1/2} &- 2\varepsilon x_1 \quad + \cdots \\
+ 1 & \quad\quad\quad = 0.
\end{aligned}
$$

At order ε^0 we are still OK as before; at order $\varepsilon^{1/2}$ we have $2x_{1/2} - 2x_{1/2} = 0$ which is also automatically satisfied. We do not get to determine anything until we go to order ε^1, where we get

$$x_{1/2}^2 + 2x_1 - 1 - 2x_1 = 0, \quad x_{1/2}^2 - 1 = 0,$$

giving two solutions $x_{1/2} = \pm 1$. Both of these are valid and will lead to valid expansions if we continue.

7.3.1.   *Choosing the expansion series*

In the example above, if we had begun by defining $\delta = \varepsilon^{1/2}$ we would have had a straightforward regular perturbation series in δ. But how do we go about spotting what series to use?

In practice, it is usually worth trying an obvious series like ε^n. But when it fails, we need a systematic method to look for the correct series.

In general, for an equation in x, we can pose a series

$$x \sim x_0 \delta_0(\varepsilon) + x_1 \delta_1(\varepsilon) + x_2 \delta_2(\varepsilon) + \cdots$$

in which x_i is strictly order 1 as $\varepsilon \to 0$ (i.e., tends neither to zero nor infinity) and the series of functions $\delta_i(\varepsilon)$ has $\delta_0(\varepsilon) \gg \delta_1(\varepsilon) \gg \delta_2(\varepsilon) \ldots$ for $\varepsilon \ll 1$. Then at each order we look for a distinguished scaling. First we look for a leading-order balance δ_0 just as in Section 7.2.1; once we have scaling δ_i we can solve for all the terms up to x_i. Substitute everything back in, tidying up the terms that cancel at each order; the next dominant scaling will give us δ_{i+1}.

If your system is particularly tricky, you may end up with logarithmic scalings for your $\delta_i(\varepsilon)$. These are difficult to deal with and result in expansions which are valid only for very small ε. See Section 1.4 of Hinch [5] for more information on log-scalings.

7.4.   *Scalings with differential equations*

7.4.1.   *Stretched coordinates*

Consider the first-order linear differential equation

$$\varepsilon \frac{\mathrm{d}f}{\mathrm{d}x} + f = 0.$$

Since it is first-order, we expect a single solution to the homogeneous equation. If we try our standard method and set $\varepsilon = 0$ we get $f = 0$ which is clearly not a good first term of an expansion!

Solving the differential equation directly gives

$$f = A_0 \exp\left[-x/\varepsilon\right].$$

This gives us the clue that what we **should** have done was change to a *stretched variable* $z = x/\varepsilon$.

Let us ignore the full solution and simply make that substitution in our governing equation. Note that $df/dx = df/dz \, dz/dx = \varepsilon^{-1}df/dz$.

$$\varepsilon\varepsilon^{-1}\frac{df}{dz} + f = 0, \qquad \frac{df}{dz} + f = 0.$$

Now the two terms balance: that is, they are the same order in ε. Clearly the solution to this equation is now $A_0 \exp[-z]$ and we have found the result.

This is a general principle. For a polynomial, we look for a distinguished scaling of the quantity we are trying to find. For a differential equation, we look for a stretched version of the independent variable.

The process is very similar to that for a polynomial. We use a trial scaling δ and set

$$x = a + \delta(\varepsilon)X.$$

Then we vary δ, looking for values at which the two largest terms in the scaled equation balance.

Let us work through the process for the following equation:

$$\varepsilon\frac{d^2 f}{dx^2} + \frac{df}{dx} - f = 0.$$

Again, we note that if $x = a + \delta X$ then $d/dx = \delta^{-1}d/dX$. We substitute in these scalings, and then look at gradually increasing δ:

$$[\mathbf{A}] \ \ \varepsilon\delta^{-2}, \quad [\mathbf{B}] \ \ \delta^{-1}, \quad [\mathbf{C}] \ \ 1.$$

For small δ, term $[\mathbf{A}]$ is the largest; as δ increases term $[\mathbf{B}]$ catches up first at $\delta = \varepsilon$. Then $[\mathbf{C}]$ catches $[\mathbf{B}]$ at $\delta = 1$ so the two distinguished stretches are $\delta = \varepsilon$ and $\delta = 1$.

For $\delta = 1$ we can treat this as a regular perturbation expansion:

$$f = f_0(x) + \varepsilon f_1(x) + \cdots,$$

$$[\varepsilon f_0''] + [f_0' + \varepsilon f_1'] - [f_0 + \varepsilon f_1] = O(\varepsilon^2),$$

Order 1: $f_0' - f_0 = 0 \qquad \Rightarrow f_0(x) = a_0 e^x,$
Order ε: $f_1' - f_1 = -a_0 e^x \ \Rightarrow \ f_1(x) = a_1 e^x - a_0 x e^x,$

so the regular solution begins

$$f(x) \sim a_0 e^x + \varepsilon(a_1 - a_0 x)e^x + \cdots.$$

For $\delta = \varepsilon$ we use our new variable $X = \varepsilon^{-1}(x - a)$ and work with the new governing equation:

$$\frac{d^2 f}{dX^2} + \frac{df}{dX} - \varepsilon f = 0.$$

Again, with the new scaling, we try a regular perturbation expansion:

$$f = f_0 + \varepsilon f_1 + \varepsilon^2 f_2 + \cdots.$$

We substitute this in and collect powers of ε:

$$\begin{aligned}
f_{0XX} + f_{0X} &= 0, \\
\varepsilon f_{1XX} + \varepsilon f_{1X} - \varepsilon f_0 &= 0, \\
\varepsilon^2 f_{2XX} + \varepsilon^2 f_{2X} - \varepsilon^2 f_1 &= 0.
\end{aligned}$$

We then solve at each order:

$$\begin{aligned}
&\varepsilon^0: f_{0XX} + f_{0X} = 0, &&f_0 = A_0 + B_0 e^{-X}, \\
&\varepsilon^1: f_{1XX} + f_{1X} - f_0 = 0, &&f_1 = A_0 X - B_0 X e^{-X} + A_1 + B_1 e^{-X}
\end{aligned}$$

and so on. At each order we introduce more unknown constants. Finally, we can express the solution in terms of our original variable:

$$\begin{aligned}
f(x) \sim A_0 + B_0 \exp\left[-\frac{(x-a)}{\varepsilon}\right] + \varepsilon \left\{ A_1 + A_0 \left(\frac{x-a}{\varepsilon}\right) \right. \\
\left. + \left(B_1 - B_0 \left(\frac{x-a}{\varepsilon}\right)\right) \exp\left[-\frac{(x-a)}{\varepsilon}\right] \right\} + \cdots.
\end{aligned}$$

Note that this expansion is only valid where $X = (x-a)/\varepsilon$ is order 1: that is, for x close to the (unknown) value a.

7.4.2. *Must two terms dominate?*

In fact we have been rather strict in our conditions. To find all roots of a polynomial, we only ever consider scalings where the two largest terms balance. But for a differential equation we can, if we like, be more relaxed. We **must** include at least one scaling in which the highest-order derivative participates, otherwise we have lost one solution of our equation; but it is possible to have a solution in which a derivative (usually the highest derivative) dominates alone. Sometimes this is a (non-fatal) sign that we could have chosen our scaling better; sometimes, in complicated systems, it's unavoidable.

Example

If we carry out a naïve regular expansion on the ODE

$$\frac{\mathrm{d}f}{\mathrm{d}x} + \varepsilon f = 0 \quad \text{with boundary condition } f(0) = C,$$

we will calculate a valid expansion of the true solution:

$$f = C\left\{1 + \varepsilon x + \tfrac{1}{2}\varepsilon x^2 + \cdots\right\}, \quad f = C\exp \varepsilon x.$$

7.5.   *Nonlinear differential equations: Scale and stretch*

For a linear differential equation, if f is a solution then so is Cf for any constant C. So if $f(x;\varepsilon)$ is a solution as an asymptotic expansion, then Cf is a valid asymptotic solution **even if C is an arbitrary function of ε.**

The same is not true of nonlinear differential equations. Suppose we are looking at the equation:

$$\frac{\mathrm{d}^2 f}{\mathrm{d}x^2} + \varepsilon f(x)\frac{\mathrm{d}f}{\mathrm{d}x} + f^2(x) = 0.$$

There are two different types of scaling we can apply: we can scale f, or we can stretch x. To get all valid scalings we need to do both of these at once.

Let us take $f = \varepsilon^\alpha F$ where F is strictly ord(1), and $x = a + \varepsilon^\beta z$ with z also strictly ord(1). Then a derivative scales like $\mathrm{d}/\mathrm{d}x \sim \varepsilon^{-\beta}\mathrm{d}/\mathrm{d}z$ and we can look at the scalings of all our terms:

$$\frac{\mathrm{d}^2 f}{\mathrm{d}x^2} \quad + \varepsilon f(x)\frac{\mathrm{d}f}{\mathrm{d}x} + f^2(x) = 0,$$

$$\varepsilon^\alpha \varepsilon^{-2\beta} \qquad \varepsilon\varepsilon^{2\alpha}\varepsilon^{-\beta} \qquad \varepsilon^{2\alpha}.$$

As always with three terms in the equation, there are three possible dominant balances.

Terms I and II: We need $\alpha - 2\beta = 2\alpha + 1 - \beta$. This gives $\alpha + \beta + 1 = 0$, so that terms I and II scale as $\varepsilon^{2+3\alpha}$, and term III scales as $\varepsilon^{2\alpha}$. We need the balancing terms to dominate, so we also need $2\alpha > 2 + 3\alpha$ which gives $\alpha < -2$.

Terms I and III: We need $\alpha - 2\beta = 2\alpha$. This gives $\alpha = -2\beta$, so that terms I and III scale as $\varepsilon^{2\alpha}$ and term II scales as $\varepsilon^{1+5\alpha/2}$. Again, we need the non-balancing term to be smaller than the others, so we need $1 + 5\alpha/2 > 2\alpha$, i.e., $\alpha > -2$.

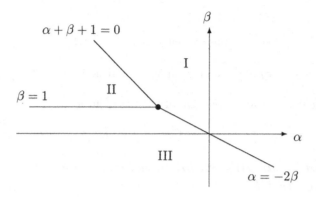

Fig. 2.   Distinguished scalings in the α–β plane for the example in Section 7.5.

Terms II and III: We need $2\alpha - \beta + 1 = 2\alpha$ which gives $\beta = 1$. Then
terms II and III scale as $\varepsilon^{2\alpha}$ and term I scales as $\varepsilon^{\alpha-2}$, so to make term
I smaller than the others we need $\alpha - 2 > 2\alpha$, giving $\alpha < -2$.

In Fig. 2, we plot the lines in the α–β plane where these balances occur,
and in the regions between, which term (I, II, or III) dominates.

   We can see that there is a distinguished scaling $\alpha = -2$, $\beta = 1$ where
all three terms balance. If we apply this scaling to have $z = (x - a)/\varepsilon$ and
$F = \varepsilon^2 f$ then the governing ODE for $F(z)$ (after multiplication of the whole
equation by ε^4) becomes

$$\frac{d^2 F}{dz^2} + F\frac{dF}{dz} + F^2 = 0.$$

This is very nice, but may not always be appropriate: often, the boundary
conditions fix the size of f or x, in which case the best you can do may be a
simple balance point (a pair (α, β) lying on one of the lines in the diagram).

7.6.   *Matching: Boundary layers*

Consider the following equation for $f(x)$ (rather similar to the example we
used in Section 7.4):

$$\varepsilon f'' + f' + f = 0.$$

There are two solutions. One is regular:

$$f = f_0(x) + \varepsilon f_1(x) + \cdots.$$

Substituting gives

Order 1: $f_0' + f_0 = 0 \quad \Rightarrow f_0 = a_0 e^{-x}$;
Order ε: $f_1' + f_1 + f_0'' = 0 \Rightarrow f_1 = [a_1 - a_0 x]e^{-x}$.

The second solution is singular, and the distinguished scaling (to balance the first two terms) is $\delta = \varepsilon$. We introduce a new variable $z = (x - a)/\varepsilon$:

$$\mathrm{d}^2 f/\mathrm{d}z^2 + \mathrm{d}f/\mathrm{d}z + \varepsilon f = 0$$

with solution

$$f = F_0(z) + \varepsilon F_1(z) + \cdots .$$

Solving, we have

Order 1: $F_0'' + F_0' = 0 \quad \Rightarrow F_0 = A_0 + B_0 e^{-z}$;
Order ε: $F_1'' + F_1' + F_0 = 0 \Rightarrow F_1 = A_1 - A_0 z + B_0[z e^{-z} + e^{-z}]$
$\qquad\qquad\qquad\qquad\qquad\quad + B_1 e^{-z}.$

We now have two possible solutions:

$$f(x) \sim a_0 e^{-x} + \varepsilon[a_1 - a_0 x]e^{-x} + \cdots ,$$
$$f(z) \sim A_0 + B_0 e^{-z} + \varepsilon[A_1 - A_0 z + B_0(z e^{-z} + e^{-z}) + B_1 e^{-z}] + \cdots .$$

Question: Will we ever need to use both of these in the same problem?
Answer: Yes. This is a second-order ODE, so we are entitled to demand that the solution satisfies two boundary conditions.

Suppose, with the differential equation above, the boundary conditions are

$$f = e^{-1} \text{ at } x = 1 \quad \text{and} \quad \mathrm{d}f/\mathrm{d}x = 0 \text{ at } x = 0.$$

We will start by assuming that the unstretched form will do, and apply the boundary condition at $x = 1$ to it:

$$f(x) \sim a_0 e^{-x} + \varepsilon[a_1 - a_0 x]e^{-x} + \cdots ,$$
$$e^{-1} = a_0 e^{-1} + \varepsilon[a_1 - a_0]e^{-1} + \cdots ,$$

which immediately yields the conditions $a_0 = 1$, $a_1 = 1$. If we had continued to higher orders we would be able to find the constants there as well.

Now what about the other boundary condition? We have no more disposable constants so we would be very lucky if it worked! In fact we have

$$\mathrm{d}f/\mathrm{d}x(x = 0) = -1 - 2\varepsilon + \cdots .$$

This is where we have to use the other solution. If we fix $a = 0$ in the scaling for z, then the strained region is near $x = 0$. We can re-express the boundary condition in terms of z:

$$\mathrm{d}f/\mathrm{d}z = 0 \text{ at } z = 0.$$

Now applying this boundary condition to our strained expansion gives:

$$f'(z) \sim -B_0 e^{-z} + \varepsilon[-A_0 - B_0 z e^{-z} - B_1 e^{-z}] + \cdots$$
$$0 \sim -B_0 + \varepsilon[-A_0 - B_1] + \cdots .$$

This fixes $B_0 = 0$, $B_1 = -A_0$ but does not determine A_0 or A_1. The solution which matches the $x = 0$ boundary condition is

$$f(z) \sim A_0 + \varepsilon[A_1 - A_0 z - A_0 e^{-z}] + \cdots .$$

We now have two perturbation expansions, one valid at $x = 1$ and for most of our region, the other valid near $x = 0$. We have not determined all our parameters. How will we do this? The answer is **matching**.

7.6.1.   *Matching: Intermediate variable*

Suppose (as in the example above) we have two asymptotic solutions to a given problem.

The outer solution scales normally and satisfies a boundary condition somewhere away from the tricky region.

The inner solution is expressed in terms of a scaled variable, and is valid in a narrow region, (probably) near the other boundary.

In order to make sure that these two expressions both belong to the same real (physical) solution to the problem, we need to **match** them.

In the case where our two solutions:

$$f_{\text{outer}}(x) = f_0(x) + \varepsilon f_1(x) + \varepsilon^2 f_2(x) + \cdots ,$$
$$f_{\text{inner}}(z) = F_0(z) + \varepsilon F_1(z) + \varepsilon^2 F_2(z) + \cdots ,$$

are linked by the scaling $z = x/\varepsilon$, we will match the two expressions using an **intermediate variable**. This is a new variable, ξ, intermediate in size between x and z, so that when ξ is order 1, x is small and z is large. We can define it as

$$x = \varepsilon^\alpha \xi \quad \Rightarrow \quad z = \varepsilon^{\alpha-1}\xi,$$

for α between 0 and 1. It is best to keep α symbolic.

The procedure is to substitute ξ into both $f(x)$ and $f(z)$ and then collect orders of ε and force the two expressions to be equal. This is best seen by revisiting the previous example.

Example continued

We had two solutions:

$$f(x) = e^{-x} + \varepsilon(1-x)e^{-x} + \cdots,$$
$$f(z) = A_0 + \varepsilon[A_1 - A_0 z - A_0 e^{-z}] + \cdots,$$

with $z = x/\varepsilon$. Defining $x = \varepsilon^\alpha \xi$, we look first at $f(x)$:

$$f(x) = e^{-\varepsilon^\alpha \xi} + \varepsilon(1 - \varepsilon^\alpha \xi)e^{-\varepsilon^\alpha \xi} + \cdots.$$

Since $\varepsilon^\alpha \ll 1$ we can expand the exponential terms to give

$$f(x) = 1 - \varepsilon^\alpha \xi - \frac{1}{2}\varepsilon^{2\alpha}\xi^2 + \varepsilon - 2\varepsilon^{\alpha+1}\xi + O(\varepsilon^2, \varepsilon^{1+2\alpha}, \varepsilon^{3\alpha}).$$

Now we look at $f(z)$. Note that $z = \varepsilon^{\alpha-1}\xi$, which is large.

$$f(z) = A_0 + \varepsilon[A_1 - A_0 \varepsilon^{\alpha-1}\xi - A_0 e^{-\varepsilon^{\alpha-1}\xi}] + \cdots.$$

Here, the exponential terms become very small indeed so we neglect them. Then comparing terms of the two expansions, we can see:

$$f(x) = 1 - \varepsilon^\alpha \xi - \tfrac{1}{2}\varepsilon^{2\alpha}\xi^2 + \varepsilon - 2\varepsilon^{\alpha+1}\xi + O(\varepsilon^2, \varepsilon^{1+2\alpha}, \varepsilon^{3\alpha}),$$
$$f(z) = A_0 - A_0\varepsilon^\alpha \xi + \varepsilon A_1 + \cdots.$$

We can match the order 1 and order ε^α terms if we set $A_0 = 1$; we can match at order ε by setting $A_1 = 1$. We have not matched the order $\varepsilon^{2\alpha}$ term, but structurally we can see that we could get this if we went to order ε^2 in the inner expansion.

We have now determined all the constants to this order: so we have the two solutions, coupled via the change of variable $x = \varepsilon z$:

$$f(x) = e^{-x} + \varepsilon(1-x)e^{-x} + \cdots,$$
$$f(z) = 1 + \varepsilon[1 - z - e^{-z}] + \cdots.$$

7.6.2. *Where is the boundary layer?*

In the last example, we **assumed** the boundary layer would be next to the lower boundary. If we did not know, how would we work it out?

Let us revisit the previous example, but attempting to put the boundary layer near $x = 1$. We will fast-forward to the matching stage, where now our intermediate variable satisfies

$$x = 1 + \varepsilon^{\alpha}\xi \quad \Rightarrow \quad z = \varepsilon^{\alpha-1}\xi$$

and ξ is negative within our domain. When ξ is order 1, z is large and negative. Recall we had an inner solution

$$F(z) \sim A_0 + B_0 e^{-z} + \varepsilon[A_1 - A_0 z + B_0(z e^{-z} + e^{-z}) + B_1 e^{-z}] + \cdots .$$

All the exponentials in this solution will be growing in the matching region! This will never match onto a well-behaved outer solution.

Key fact: The boundary layer is always positioned so that any exponentials in the inner solution **decay** as you move towards the outer.

8. Exercises

(1) Find the general solution to the PDE for $f(\theta, \phi)$:

$$\frac{1}{a\sin\theta}\frac{\partial}{\partial\theta}(\sin\theta v_\theta f) + \frac{1}{a\sin\theta}\frac{\partial}{\partial\phi}(v_\phi f) + \sin^2\theta\cos 2\phi = 0$$

in which $v_\theta = a\sin\theta\cos\theta\cos 2\phi$ and $v_\phi = -a\sin\theta\sin 2\phi$.

(2) Consider the problem

$$\frac{\partial u}{\partial t} + u^2\frac{\partial u}{\partial x} = 0,$$

in $x \geq 0$, $t \geq 0$, with initial and boundary conditions

$$u(x,0) = \sqrt{x} \quad u(0,t) = 0.$$

Find the general solution implicitly and hence the specific solution.

(3) Solve the following PDE with the boundary conditions given:

$$\frac{\partial^2 u}{\partial t^2} - \frac{x^2}{(t+1)^2}\frac{\partial^2 u}{\partial x^2} = 0, \quad u(x,0) = u(1,t) = u(2,t) = 0.$$

(4) Try a regular perturbation expansion in the following ODE:
$$y'' + 2\varepsilon y' + (1 + \varepsilon^2)y = 1, \quad y(0) = 0, \quad y(\pi/2) = 0.$$
Calculate y correct to order ε^2.

(5) Find two terms of a regular perturbation expansion for $f(x,t)$ in:
$$\frac{\partial^2 f}{\partial t^2} - \frac{\partial^2 f}{\partial x^2} - \varepsilon \cos x f = x,$$
with boundary conditions $f(x,0) = \partial f/\partial t(x,0) = 0$.

(6) Find the first two terms of all four roots of $\varepsilon x^4 - x^2 - x + 2 = 0$.

(7) Consider the following ODE for $f(x)$: $\varepsilon f'' + f f' - f = 0$.

  (a) Find the scalings $f = \varepsilon^\alpha F$ and stretches $x = a + \varepsilon^\beta z$ at which two dominant terms balance, and sketch these scalings in the α–β plane.

  (b) Hence, determine the critical α and β for all three terms to balance.

  (c) Give also the possible values of β if the boundary conditions fix $\alpha = 0$. Find the leading term in an expansion for f in each case.

(8) Find the distinguished stretches, and the leading term of each solution:
$$\varepsilon^3 \frac{d^3 f}{dx^3} + \varepsilon \frac{d^2 f}{dx^2} + \frac{df}{dx} + f = 0.$$

(9) Use matched asymptotic expansions to solve the following system:
$$\frac{x}{y} \frac{\partial f}{\partial x} - \frac{\partial f}{\partial y} + \frac{f}{y} - \varepsilon \nabla^2 f = 0,$$
$$f + \varepsilon \partial f/\partial y = 0 \ \text{ at } \ y = 1, \qquad f = 2 \ \text{ at } \ y = 2.$$

(10) Consider the following equation and boundary conditions:
$$\varepsilon \frac{\partial^2 u}{\partial x^2} + \varepsilon \frac{\partial^2 u}{\partial y^2} + \frac{\partial u}{\partial y} = 0,$$
$$u(-1, y) = u(1, y) = 0, \quad u(x, 1) = 1 - x^2,$$
$$\varepsilon \frac{\partial u}{\partial y}(x, 0) + u(x, 0) = 0.$$

  (a) Calculate the first two terms of a perturbation expansion for u, using matched asymptotic expansions.

  (b) Taking ε as a normal parameter (i.e., forgetting that it is small), find the full solution to the problem by separating variables.

  (c) Comment on the structure of your general solution when ε is small.

9. Worked solution to example 9

9.1. *Outer: Leading-order term*

We expand the PDE:

$$\frac{x}{y}\frac{\partial f}{\partial x} - \frac{\partial f}{\partial y} + \frac{f}{y} - \varepsilon\frac{\partial^2 f}{\partial x^2} - \varepsilon\frac{\partial^2 f}{\partial y^2} = 0$$

and look for the first term of an outer solution by considering the case $\varepsilon = 0$:

$$\frac{x}{y}\frac{\partial f}{\partial x} - \frac{\partial f}{\partial y} + \frac{f}{y} = 0, \quad x\frac{\partial f}{\partial x} - y\frac{\partial f}{\partial y} + f = 0.$$

Because this is a first-order PDE we can apply the method of characteristics. Solving for the characteristic curves:

$$\partial x/\partial r = x, \quad \partial y/\partial r = -y,$$

which gives us the parametric curves $x = x_0 e^r$, $y = e^{-r}$ along which

$$\frac{\partial f}{\partial r} = -f, \quad f_0 = A(x_0)e^{-r} = A(xy)y.$$

We will stay at one term for the outer solution for now.

9.2. *Inner: Scaling and leading-order term*

We are expecting a boundary layer somewhere, because all the highest derivatives were neglected when we put $\varepsilon = 0$. Since boundary layers usually live near boundaries, and there are no boundaries on x, we choose to scale y; another hint is the form of the boundary condition at $y = 1$.

We set $y = a + \varepsilon^b z$ and substitute into the PDE:

$$\frac{x}{[a + O(\varepsilon^b)]}\frac{\partial f}{\partial x} - \varepsilon^{-b}\frac{\partial f}{\partial z} + \frac{f}{[a + O(\varepsilon^b)]} - \varepsilon\frac{\partial^2 f}{\partial x^2} - \varepsilon^{1-2b}\frac{\partial^2 f}{\partial z^2} = 0.$$

Clearly the two terms which may be larger than $O(1)$ if $b > 0$ are the second and last terms: ε^{-b} and ε^{1-2b}. Balancing the two fixes $b = 1$ (which we expected from the boundary condition). Thus:

$$-\frac{\partial f}{\partial z} - \frac{\partial^2 f}{\partial z^2} + \varepsilon\frac{x}{a}\frac{\partial f}{\partial x} + \varepsilon\frac{f}{a} = O(\varepsilon^2).$$

Let us just look at the leading-order term first: $f = f_0 + \varepsilon f_1 + \cdots$ gives

$$-\partial f_0/\partial z - \partial^2 f_0/\partial z^2 = 0, \quad f_0 = A_0(x) + B_0(x)e^{-z}.$$

The sign of the exponential in z tells us that the boundary layer must be located at the lower boundary so we fix $a = 1$ and $y = 1 + \varepsilon z$. We expect the outer to satisfy the upper boundary condition at $y = 2$.

9.3. *Outer: Full solution*

We now have the outer solution

$$f = A(xy)y + \varepsilon f_1(x,y) + \varepsilon^2 f_2(x,y) + \cdots ,$$

which we need to satisfy the boundary condition $f(x,2) = 2$. Applying this at leading-order gives

$$2 = 2A(2x), \quad A(\eta) = 1, \quad f = y + \varepsilon f_1(x,y) + \varepsilon^2 f_2(x,y) + \cdots .$$

Now we can continue with the expansion: the original equation was

$$x\frac{\partial f}{\partial x} - y\frac{\partial f}{\partial y} + f - y\varepsilon\frac{\partial^2 f}{\partial x^2} - y\varepsilon\frac{\partial^2 f}{\partial y^2} = 0,$$

so we have

$$\begin{aligned}
x\partial f_0/\partial x - y\partial f_0/\partial y + f_0 &= 0, \\
x\partial f_1/\partial x - y\partial f_1/\partial y + f_1 - y\partial^2 f_0/\partial x^2 - y\partial^2 f_0/\partial y^2 &= 0, \\
x\partial f_2/\partial x - y\partial f_2/\partial y + f_2 - y\partial^2 f_1/\partial x^2 - y\partial^2 f_1/\partial y^2 &= 0,
\end{aligned}$$

with boundary conditions

$$f_0(x,2) = 2, \quad f_1(x,2) = 0, \quad f_2(x,2) = 0.$$

At order 1 we know this is satisfied by $f_0 = y$. At order ε we have

$$x\frac{\partial f_1}{\partial x} - y\frac{\partial f_1}{\partial y} + f_1 = 0,$$

which is the same equation we had for f_0, so has solution $f_1 = A_1(xy)y$. This time the boundary condition gives $f_1 = 0$. We can see that this pattern will continue, and in fact $f_n = 0$ for $n \geq 1$: the full outer solution is

$$f_{\text{outer}} = y.$$

9.4. *Inner: First two terms of the expansion*

We now return to the inner solution:

$$\frac{\partial^2 f}{\partial z^2} + \frac{\partial f}{\partial z} = \varepsilon \frac{x}{[1+\varepsilon z]}\frac{\partial f}{\partial x} + \varepsilon\frac{f}{[1+\varepsilon z]} - \varepsilon^2 \frac{\partial^2 f}{\partial x^2}.$$

Keeping terms up to order ε gives

$$\frac{\partial^2 f_0}{\partial z^2} + \frac{\partial f_0}{\partial z} = 0; \quad \frac{\partial^2 f_1}{\partial z^2} + \frac{\partial f_1}{\partial z} = x\frac{\partial f_0}{\partial x} + f_0,$$

with boundary conditions (true at each order) $f + \partial f/\partial z = 0$ at $z = 0$.

Leading-order $f_0 = A_0(x) + B_0(x)e^{-z}$, and the boundary condition gives $A_0(x) = 0$: $f_0 = B_0(x)e^{-z}$.

First-order At order ε, we have

$$\frac{\partial^2 f_1}{\partial z^2} + \frac{\partial f_1}{\partial z} = [xB_0' + B_0]e^{-z},$$

which gives $f_1 = A_1(x) + B_1(x)e^{-z} - [xB_0' + B_0]ze^{-z}$. Applying the boundary condition fixes $A_1(x) = [xB_0' + B_0]$.

9.5. *Matching*

Our two solutions are $f_{\text{outer}} = y$ and

$$f_{\text{inner}} = B_0(x)e^{-z} + \varepsilon[(xB_0'(x) + B_0(x))(1 - ze^{-z}) + B_1(x)e^{-z}] + O(\varepsilon^2).$$

Using an intermediate variable $y = 1 + \varepsilon^{\alpha}\eta$, $z = \varepsilon^{\alpha-1}\eta$, these become

$$f_{\text{outer}} = 1 + \varepsilon^{\alpha}\eta; \quad f_{\text{inner}} = \varepsilon(xB_0'(x) + B_0(x)) + O(\varepsilon^2).$$

There is nothing in the inner large enough to match onto the 1 in the outer.

However, remember we started from a **linear** equation. Along with the fact that the inner boundary condition was homogeneous, that means that if f_{inner} is a solution, so is $\varepsilon^{-1} f_{\text{inner}}$. So we try that:

$$f_{\text{inner,new}} = \varepsilon^{-1} B_0(x) e^{-z} + (x B_0'(x) + B_0(x))(1 - z e^{-z}) + B_1(x) e^{-z} + O(\varepsilon)$$
$$\sim (x B_0'(x) + B_0(x)) + O(\varepsilon) \quad \text{as } z \to \infty.$$

Now we can match the two functions if

$$x B_0'(x) + B_0(x) = 1,$$

which is just an ODE. Solving gives $B_0(x) = 1 + C/x$ and since the line $x = 0$ is within our domain, we require $C = 0$ for regularity. Thus:

$$f_{\text{outer}} = y,$$
$$f_{\text{inner}} = \varepsilon^{-1} e^{-z} + [(1 - z e^{-z}) + B_1(x) e^{-z}] + O(\varepsilon),$$

with $y = 1 + \varepsilon z$. To determine B_1 we would have to calculate the f_2 term of the inner expansion.

References

[1] H. F. Weinberger, *A First Course in Partial Differential Equations with Complex Variables and Transform Methods*. Xerox College Publishing, New York, 1965.

[2] C. M. Bender and S. A. Orszag, *Advanced Mathematical Methods for Scientists and Engineers*. Springer, New York, 1999.

[3] J. Kevorkian and J. D. Cole, *Perturbation Methods in Applied Mathematics*. Springer-Verlag, New York, 1985.

[4] M. D. Van Dyke, *Perturbation Methods in Fluid Mechanics*. Parabolic Pr., Stanford, California, 1975.

[5] E. J. Hinch, *Perturbation Methods*. Cambridge University Press, Cambridge, 1991.

[6] J. W. Brown and R. V. Churchill, *Complex Variables and Applications*. McGraw-Hill, New York, 2009.

[7] Z. Nehari, *Introduction to Complex Analysis*. Literary Licensing, LLC, 2013.

Chapter 2

Resonances in Wave Scattering

Dmitry V. Savin

Department of Mathematics, Brunel University London,
Uxbridge UB8 3PH, UK

This chapter provides an introduction into the theory of resonant scattering of waves, beginning with classical fields and going over to nonrelativistic quantum scattering. Basic concepts of resonances are first introduced using model examples and then extended to a more general setting. An emphasis is placed on the connection between causality and analyticity of scattering amplitudes, and on dispersion relations arising.

1. Introduction

Resonance phenomena have their historical roots in acoustics and mechanical vibrations, with numerous applications ranging from electromagnetism and optics to quantum mechanics and particle physics. Many of their key properties can be described within the classical wave equation. Section 2 provides a pedagogical primer illustrating the main features of resonant scattering on an example of a linear oscillator excited by a driving force. The causal character and related analytic properties of the response in general linear systems are then discussed in Section 3 on the basis of Titchmarsh's theorem. This is further used in Section 4 to construct the canonical product expansion of the S matrix for classical waves in terms of its poles (resonances), as dictated by causality. Finally, we carry out in Section 5 a similar program in the case of quantum scattering in cutoff potentials, highlighting similarities and important differences that arise.

Most of the chapter follows the excellent monograph by Nussenzveig [1], the full text of which is also available online through Elsevier's series *Mathematics in Science and Engineering*, Vol. 95. Here, we consider scattering in one dimension, which is sufficient to discuss all the essential concepts, referring to Nussenzveig's text for further generalities. The monograph by Perelomov and Zeldovich [2] is recommended for advanced study of resonance scattering in quantum mechanics. Taylor's textbook [3] gives a general description of the scattering theory as a practical tool in physics. For more formal mathematical scattering theory the interested reader is referred to the texts by Reed and Simon [4] and Yafaev [5].

2. Linear classical oscillator

The key characteristics of resonance phenomena can be illustrated on a model example of a harmonic oscillator. It also provides a useful basis for studying general linear systems through the technique of Fourier analysis.

2.1. *Simple versus damped harmonic motion*

Simple harmonic motion (SHM) corresponds to periodic motion of a point mass m around the equilibrium position (say, at $x = 0$) subject to a linear restoring force. The equation of motion reads

$$\ddot{x} + \omega_0^2 x = 0, \tag{1}$$

where dots denote time derivatives and ω_0 is the natural frequency of the oscillations. The real general solution of (1) can be written as

$$x(t) = a\cos(\omega_0 t + \phi) = Ae^{-i\omega_0 t} + Be^{i\omega_0 t}, \tag{2}$$

with complex $B^* = A = (a/2)e^{-i\phi}$. Two real parameters, the amplitude a and initial phase ϕ, are determined from the initial conditions. The frequency spectrum of SHM is therefore *discrete* and consists of two lines at $\pm\omega_0$. This type of motion describes *stationary* oscillations for which the total energy remains constant at all times (i.e., the integral of motion)

$$E = \frac{m\dot{x}^2}{2} + \frac{m\omega_0^2 x^2}{2} = \frac{m}{2}\omega_0^2 a^2. \tag{3}$$

Damped harmonic motion (DHM) occurs due to an interaction with the environment leading to energy dissipation. Assuming the dissipative force $-\gamma m v$, with damping $\gamma > 0$, the equation of motion takes the form

$$\ddot{x} + \gamma\dot{x} + \omega_0^2 x = 0. \tag{4}$$

It has a general solution as a linear superposition $x(t) = Ae^{-i\omega_1 t} + Be^{-i\omega_2 t}$, where $\omega_{1,2}$ are now the (complex) roots of the characteristic equation

$$\omega^2 + i\gamma\omega - \omega_0^2 \equiv (\omega - \omega_1)(\omega - \omega_2) = 0, \tag{5}$$

i.e., $\omega_{1,2} = \pm\sqrt{\omega_0^2 - \gamma^2/4} - i\gamma/2 \equiv \pm\widetilde{\omega}_0 - i\gamma/2$. Note here the always negative imaginary parts and also the important symmetry relation $\omega_2 = -\omega^*$.

When excited at $t = 0$, the motion then evolves according to $(t \geq 0)$

$$x(t) = (Ae^{-i\widetilde{\omega}_0 t} + Be^{i\widetilde{\omega}_0 t})e^{-\gamma t/2} = ae^{-\gamma t/2}\cos(\widetilde{\omega}_0 t + \phi), \tag{6}$$

which describes damped oscillations and features two new effects. Firstly, there is a shift in the frequency or period of the free oscillations, $T = 2\pi/\omega_0$ (this can however be neglected if dissipation is small). Secondly and more importantly, there appears an overall exponential decay of the oscillations with time (hence, the total energy $\propto e^{-\gamma t}$), which persists at any finite γ. The latter gives rise to a new time scale, the 'lifetime' of the oscillations $\tau = 1/\gamma$. In the most interesting case of $\tau \gg T$ or $\gamma \ll \omega_0$, the motion undergoes many $(\sim \tau/T \sim \omega_0/\gamma \gg 1)$ oscillations before a noticeable decay of their amplitude (or energy). Such a case corresponds to the regime of the so-called *quasistationary* oscillations. (Note that $\widetilde{\omega}_0$ becomes imaginary in the opposite case $\gamma \gg \omega_0$, the motion being completely aperiodic.)

The spectrum of the damped oscillations is no longer discrete and becomes *continuous*. To see this, let us represent $x(t)$ as a Fourier integral

$$x(t) = \frac{1}{2\pi}\int_{-\infty}^{\infty} X(\omega)e^{-i\omega t}d\omega, \qquad X(\omega) = \int_{-\infty}^{\infty} x(t')e^{i\omega t'}dt'. \tag{7}$$

Evaluating $X(\omega) = (a/2)\int_0^{\infty}[e^{i(\omega-\omega_1)t'} + e^{i(\omega-\omega_2)t'}]dt'$ (we put here $\phi = 0$ for simplicity) and making explicit use of (5), we arrive at

$$X(\omega) = ia\frac{\omega + i\gamma/2}{(\omega - \omega_1)(\omega + \omega_1^*)} = ia\frac{\omega + i\gamma/2}{\omega^2 + i\gamma\omega - \omega_0^2}. \tag{8}$$

Hence, the spectral intensity of the damped oscillations is nonzero at all frequencies, $|X(\omega)|^2 = a^2(\omega^2 + \gamma^2/4)/[(\omega^2 - \omega_0^2)^2 + \gamma^2\omega^2]$. In the regime of quasistationary oscillations, $\gamma \ll \omega_0$, the intensity exhibits a strong peak near $\omega \approx \pm\omega_0$, where it simplifies to the following expression

$$\frac{|X(\omega)|^2}{|X(\pm\omega_0)|^2} \approx \frac{\gamma^2/4}{(\omega \mp \omega_0)^2 + \gamma^2/4} \equiv \mathcal{I}_{\pm\omega_0}(\omega). \tag{9}$$

Such a bell-shape curve is very common for resonance effects and often called the *Breit–Wigner formula* (nuclear and particle physics) or

Cauchy–Lorentz distribution (mathematics). It is centred at the resonance position, and has a half width at half maximum equal to $\gamma/2 = 1/2\tau$. Hence, a typical spectral width $\Delta\omega \sim \gamma$ of the resonance is due to the finite lifetime, implying the relation $\Delta\omega\,\tau \sim 1$. This can be treated in the spirit of the *uncertainty relation* showing that any process with finite duration cannot be monochromatic: the bigger its spectral width is the faster it decays.

2.2. *Stationary solution and resonance*

Free oscillations decay with time and have to be sustained by an external driving force, $\mathcal{F}(t)$. Consider first the case of a harmonic $f(t) = \mathcal{F}(t)/m$,

$$\ddot{x} + \gamma\dot{x} + \omega_0^2 x = f(t) \equiv \mathrm{Re}(F_\omega e^{-i\omega t}). \tag{10}$$

We look for the so-called *response*, which is a particular solution of (10) that vanishes as $F_\omega \to 0$. It has a harmonic form with the same frequency,

$$x(t) = \mathrm{Re}(X_\omega e^{-i\omega t}), \qquad X_\omega = -\frac{F_\omega}{\omega^2 + i\gamma\omega - \omega_0^2} \equiv G(\omega)F_\omega. \tag{11}$$

The explicit expression of the stationary solution is readily found to be

$$x(t) = -\frac{|F_\omega|\cos(\omega t - \delta_\omega)}{\sqrt{(\omega^2 - \omega_0^2)^2 + \gamma^2\omega^2}}, \qquad \tan\delta_\omega = -\frac{\gamma\omega}{\omega^2 - \omega_0^2}, \tag{12}$$

where δ_ω is the *phase shift* between the response and the driving force.

For such stationary oscillations the mean total energy $m\dot{x}^2/2 + m\omega_0^2 x^2/2$ integrated over the period is constant (at given ω) and equal to $E(\omega) = (m|F_\omega|^2/4)(\omega^2 + \omega_0^2)/[(\omega^2 - \omega_0^2)^2 + \gamma^2\omega^2]$. It attains the maximum value of $E(\omega_0) = m|F_\omega|^2/2\gamma^2$ exactly at the natural frequency, vanishing quickly as $|\omega - \omega_0|$ grows. Therefore, even a weak force can cause a strong (resonant) response if damping is small enough. At $\gamma \ll \omega_0$, the energy distribution shows rapid variations when the driving frequency $\omega \approx \omega_0$, according to

$$E(\omega) \approx \frac{\gamma^2/4}{(\omega - \omega_0)^2 + \gamma^2/4} E(\omega_0) \equiv \mathcal{I}_{\omega_0}(\omega)E(\omega_0). \tag{13}$$

Thus the stationary response has the same resonance profile as the spectral intensity (9) of free damped oscillations, with the spectral width being given by the damping rate γ. Note that in the vicinity of the resonance, $\omega \sim \omega_0$, the phase shift has the following simple form

$$\delta_\omega \approx -\tan^{-1}\frac{\gamma/2}{\omega - \omega_0}. \tag{14}$$

It undergoes a sharp change from 0 to π, going through the value of $\pi/2$ at $\omega = \omega_0$. Hence, the resonance parameters can be obtained from δ_ω.

2.3. *General solution, causality, and analyticity*

The response to an arbitrary driving force can be found by taking into account the *linear* nature of the problem. Since we can represent the driving force by a Fourier integral, $f(t) = \frac{1}{2\pi} \int_{-\infty}^{\infty} F(\omega)e^{-i\omega t}d\omega$, the resulting response is equal to the *superposition* of all the harmonic terms

$$x(t) = \frac{1}{2\pi} \int_{-\infty}^{\infty} X(\omega)e^{-i\omega t}d\omega \equiv \frac{1}{2\pi} \int_{-\infty}^{\infty} G(\omega)F(\omega)e^{-i\omega t}d\omega. \qquad (15)$$

Substituting $F(\omega) = \int_{-\infty}^{\infty} f(t')e^{i\omega t'} dt'$ and inverting the order of integration (assuming that all functions are sufficiently well behaved), we arrive at the well-known representation in terms of a convolution product

$$x(t) = \int_{-\infty}^{\infty} g(t-t')f(t')dt', \qquad g(\tau) \equiv \frac{1}{2\pi} \int_{-\infty}^{\infty} G(\omega)e^{-i\omega\tau}d\omega. \qquad (16)$$

This expression describes the motion of the oscillator due to the action of the driving force, being thus determined by the function $G(\omega)$, see (11),

$$G(\omega) = -\frac{1}{(\omega - \omega_1)(\omega + \omega_1^*)}. \qquad (17)$$

Its time-domain 'counterpart' $g(t)$ is, by construction, the solution of (10) corresponding to $f(t) = \delta(t)$ (Dirac's delta-function), i.e., *Green's function*. The latter plays a fundamental role here and requires further discussion.

To get an explicit expression for $g(\tau)$, we evaluate the integral in (16) using Cauchy's residue theorem. The important point is that the poles of $G(\omega)$ always lie in the lower half of the complex ω-plane. We will use the notation I_+ (I_-) for the upper (lower) half-plane, and I_0 for the real axis. If $\tau < 0$, the path of integration can be closed in I_+, where $G(\omega)$ is regular, thus $g(\tau < 0) \equiv 0$. If $\tau > 0$, we can close the path of integration in I_- and compute the sum of two residues at ω_1 and $-\omega_1^*$, finding

$$g(\tau) = \frac{\sin(\widetilde{\omega}_0\tau)}{\widetilde{\omega}_0}e^{-\gamma\tau/2} \qquad (\tau > 0). \qquad (18)$$

Comparing with (6) gives the physical interpretation of the above result: an instantaneous 'kick' of the oscillator excites its free modes of oscillations.

This example clearly demonstrates the *causal* nature of the response: *the "effect" cannot precede the "cause"*, $g(\tau) = 0$ at $\tau < 0$. On the other hand, it also reveals the important connection with the analytic properties of $G(\omega)$ that *has a regular analytic continuation in I_+.*

2.4.   *Toy model: Vibrating string coupled to oscillator*

Consider an initial-value problem for a semi-infinite vibrating string coupled to a harmonic oscillator. This simple model brings out the main features of the general method, showing also that both damping and driving forces have the same source due to coupling the system to the radiation field [1, 6].

Let $y(x,t)$ denote the transversal displacement of the string that has the rest position coinciding with the positive x-axis. The oscillator, located at the origin, can move only vertically and its displacement is given by $y(0,t)$. The tension σ of the string exerts the force $\sigma \partial y(0,t)/\partial x$ on the oscillator, thus its equation of motion [cf. (1)] acquires a nonzero right-hand side

$$\left(\frac{\partial^2}{\partial t^2} + \omega_0^2\right) y(0,t) = \frac{\sigma}{m}\frac{\partial y(0,t)}{\partial x}. \tag{19}$$

This may be regarded as a boundary condition to the string motion equation

$$\frac{\partial^2 y}{\partial t^2} - c^2\frac{\partial^2 y}{\partial x^2} = 0 \qquad (x > 0), \tag{20}$$

which is simply the wave equation with velocity c. The general solution of (20) can be written in terms of incoming and outgoing waves as

$$y(x,t) = \varphi_{\text{in}}(t + x/c) + \varphi_{\text{out}}(t - x/c). \tag{21}$$

So, there are many regimes of motion depending on the choice of the initial conditions for the string and oscillator. We discuss the following two.

Radiation damping. At $t = 0$, the string is at rest but the oscillator is not. At $t > 0$, it excites only the outgoing wave $\varphi_{\text{out}}(t - x/c)$. Assuming also that there is no source of waves at infinity, we get an additional condition

$$\varphi_{\text{in}}(t + x/c) = 0. \tag{22}$$

Since $\partial y(x,t)/\partial x|_{x=0} = -c^{-1}\partial\varphi_{\text{out}}(t)/\partial t$, one immediately sees from (19) that the oscillator displacement $y(0,t) = \varphi_{\text{out}}(t)$ satisfies exactly the DHM equation (4), where $\gamma = \sigma/mc$. The damping is now caused by transferring the oscillator energy into the outgoing wave (radiation). This leads to the natural spectral profile (9) characterised by the radiation width γ.

Resonance scattering. Consider now the process of scattering when an incident wave gets absorbed and re-emitted by the oscillator in the opposite direction. The stationary ($\propto e^{-i\omega t}$) solution of (19) and (20) is given by a linear superposition of the plane monochromatic waves $e^{-i\omega(t\pm x/c)}$ travelling in the opposite directions, and can be represented as

$$y_\omega(x,t) = A_\omega[e^{-i\omega x/c} - S(\omega)e^{i\omega x/c}]e^{-i\omega t}. \tag{23}$$

A_ω and S are fixed by the boundary conditions at $x = 0$ and $x \to \infty$. The quantity $|A_\omega|^2$ has the meaning of the intensity of the incident wave (which has the source at infinity). Requiring $y_\omega(0,t)$ to satisfy (19) yields

$$S(\omega) = \frac{\omega^2 - i\gamma\omega - \omega_0^2}{\omega^2 + i\gamma\omega - \omega_0^2} = \frac{(\omega - \omega_1^*)(\omega + \omega_1)}{(\omega - \omega_1)(\omega + \omega_1^*)}, \tag{24}$$

where we have used the roots of the characteristic equation (5).

Function $S(\omega)$ plays the central role in scattering and is called the *scattering function* (matrix, in more general case). As is clear from (24), the S function is unimodular, $|S(\omega)| = 1$ at real ω, which expresses the energy conservation in the scattering process (the same intensities of the incident and outgoing waves, $|A_\omega|^2 = |A_\omega S(\omega)|^2$). Thus it can be written as

$$S(\omega) = e^{2i\delta(\omega)}, \qquad \tan\delta(\omega) = -\frac{\gamma\omega}{\omega^2 - \omega_0^2}. \tag{25}$$

Therefore, the *scattering phase* $\delta(\omega)$ fully describes the result of scattering on the oscillator. Note that $\delta(\omega)$ has exactly the same form as the phase shift (12), since the incident wave now acts as the effective 'driving' force.

It is also useful to introduce the so-called *scattering amplitude*

$$T(\omega) \equiv -i[1 - S(\omega)] = -2e^{i\delta(\omega)}\sin\delta(\omega). \tag{26}$$

Its magnitude squared (called a cross-section), $|T(\omega)|^2$, quantifies the intensity of excitation. In particular, the amplitude of the oscillator displacement is found from (23) and (26) as $y_\omega(0,t) = iA_\omega T(\omega)e^{-i\omega t}$, where

$$T(\omega) = \frac{2\gamma\omega}{\omega^2 + i\gamma\omega - \omega_0^2} \approx \frac{\gamma}{\omega - \omega_0 + i\gamma/2} \quad (\omega \approx \omega_0). \tag{27}$$

The last expression here stands for the Breit–Wigner approximation near ω_0, and corresponds to the resonance profile (9) for the intensity.

It is instructive to look at the analytic properties of $S(\omega)$. Function (24) has two resonance poles $(\omega_1, -\omega_1^*)$ located symmetrically with respect to the imaginary axis in I_-. The reflected image $(-\omega_1, \omega_1^*)$ in I_+ provides the zeros. For this example, we have already discussed that such a structure is consistent with the causality and unitarity conditions of $S(\omega)$. We will now see that the representation similar to (24) can be derived for rather general linear systems by requiring these two conditions alone.

3.   Causal transform

Let us consider a response ("output") to a time-dependent excitation ("input") in an arbitrary *linear* system. The superposition principle requires the output to be a linear functional of the input. Most of physical systems also possess *time-translation invariance*, so that any time shifted input results in the output shifted by the same time interval. For such systems the input–output relation can then be represented in the general form (16), where $x(t)$ denotes the output and $f(t)$ the input. We now impose that the integral kernel $g(t - t')$, which describes the system response to a 'kick' [cf. (18)], vanishes identically at $t < t'$ (causality) and is also square integrable (ensuring finite total energy). Then the function

$$G(\omega) \equiv \int_0^\infty g(\tau)e^{i\omega\tau}\,d\tau, \qquad \int_{-\infty}^\infty |G(\omega)|^2 d\omega < \infty, \qquad (28)$$

is called a *causal transform* and satisfies the following theorem.

3.1.   *Titchmarsh's theorem*

Theorem [1, Theorem 1.6.1]. *If a square integrable function $G(\omega)$ fulfills one of the four conditions below, then it fulfills all four of them:*

(i) *The inverse Fourier transform (FT) $g(t)$ of $G(\omega)$ vanishes for $t < 0$.*

(ii) *$G(u)$ is (for almost all u) the limit as $v \to 0^+$ of an analytic function $G(u+iv)$ that is holomorphic in I_+ and square integrable over any line parallel to the real axis ($v > 0$): $\int_{-\infty}^\infty |G(u + iv)|^2 du < \infty$.*

(iii) *Re G and Im G verify the first Plemelj formula:*

$$\mathrm{Re}G(\omega) = \frac{1}{\pi}\mathcal{P}\int_{-\infty}^\infty \frac{\mathrm{Im}G(\omega')}{\omega' - \omega}d\omega'.$$

(iv) *Im G and Re G verify the second Plemelj formula:*

$$\mathrm{Im}\,G(\omega) = -\frac{1}{\pi}\mathcal{P}\int_{-\infty}^\infty \frac{\mathrm{Re}\,G(\omega')}{\omega' - \omega}d\omega'.$$

A simple example of a causal transform is $G(\omega) = 1/(\omega - \zeta)$, with $\mathrm{Im}\zeta < 0$. However, we note that $G(\omega) = e^{-i\omega a}/(\omega - \zeta)$ ($\mathrm{Im}\zeta < 0$, $a > 0$) is not a causal transform, as its inverse FT is equal to zero only for $t < -a$. Also, the factor $e^{-i\omega a}$ blows up exponentially in I_+, so (ii) is not fulfilled either.

We sketch the essential steps of the proof, referring for details to Ref. [1]. First, note that $G(\omega)$ has a regular analytic continuation in I_+, which is ensured by the fact that the integral in (28) extends only to the positive times. Step (i)→(ii) follows then by virtue of Parseval's formula [7]

$$\int_{-\infty}^{\infty} f(t)g^*(t)dt = \frac{1}{2\pi}\int_{-\infty}^{\infty} F(\omega)G^*(\omega)d\omega, \qquad (29)$$

which relates the FTs $F(\omega)$ and $G(\omega)$ of respective functions $f(t)$ and $g(t)$. Thus $\int_{-\infty}^{\infty} |G(u+iv)|^2 du = 2\pi \int_0^{\infty} e^{-2vt}|g(t)|^2 dt < 2\pi \int_0^{\infty} |g(t)|^2 dt < \infty$.

To show that (ii) implies (iii) and (iv), we take a point $\omega = u_0 + iv_0$ in I_+ and apply Cauchy's theorem to $G(\omega)$, considering a rectangular contour of integration Γ with the corners at $\pm U$, $\pm U + iV$. In the limit $U, V \to \infty$, we find (as the integrals over the right, left and upper sides go to zero)

$$G(\omega) = \frac{1}{2\pi i}\int_{\Gamma} \frac{G(\omega')}{\omega' - \omega}d\omega' = \frac{1}{2\pi i}\int_{-\infty}^{\infty} \frac{G(\omega')}{\omega' - \omega}d\omega' \qquad (\operatorname{Im}\omega > 0). \quad (30)$$

This provides analytic continuation in I_+ in terms of values at I_0.

Now let ω lie on the real axis. In this case, the contour Γ must be taken to avoid the point ω by a semicircle of infinitesimal radius in I_+. This leads to a well-known expression involving the principle value ($\mathcal{P}$) integral [7]

$$0 = \frac{1}{2\pi i}\int_{\Gamma} \frac{G(\omega')}{\omega' - \omega}d\omega' = \frac{1}{2\pi i}\mathcal{P}\int_{-\infty}^{\infty} \frac{G(\omega')}{\omega' - \omega}d\omega' - \frac{G(\omega)}{2},$$

which yields the desired representation on the real axis

$$G(\omega) = \frac{1}{\pi i}\mathcal{P}\int_{-\infty}^{\infty} \frac{G(\omega')}{\omega' - \omega}d\omega' \qquad (\text{real }\omega). \qquad (31)$$

Taking the real and imaginary parts, one verifies both Plemelj's formulae. These relations between Re G and Im G are called *the dispersion relations*.

   Example. Let Im $G(\omega) = -1/(1+\omega^2)$. Then the real part is given by Re $G(\omega) = (1/\pi)\mathcal{P}\int_{-\infty}^{\infty}[(\omega'+i)(i-\omega')(\omega'-\omega)]^{-1}d\omega' = \omega/(1+\omega^2)$. This leads to $G(\omega) = 1/(\omega+i)$. Thus, the dispersion relations allow us to restore the whole analytic function only if its imaginary (or real) part is known.

3.2.   *Dispersion relations with subtraction*

Very often, the function $G(\omega)$ may not verify the assumption of square integrability and turns out to be only bounded, $|G(\omega)|^2 \leq C < \infty$. Still, $G(\omega)$ has a regular analytic continuation in I_+ but, because of the weaker condition than in (28), $G(\omega)$ itself need not be a causal transform (so Titchmarsh's theorem does not apply). To rectify the situation, one usually considers a *subtraction* at some (real) frequency $\omega = \omega_\star$,

$$\widetilde{G}(\omega) = [G(\omega) - G(\omega_\star)]/(\omega - \omega_\star).$$

By construction, $\widetilde{G}(\omega)$ is bounded (assuming that $dG/d\omega|_{\omega_\star}$ exists), regular in I_+, and also square integrable on the real axis. We may then apply Titchmarsh's theorem, thus (31), to the subtraction $\widetilde{G}(\omega)$ and obtain

$$G(\omega) = G(\omega_\star) + \frac{(\omega - \omega_\star)}{\pi i} \mathcal{P} \int_{-\infty}^{\infty} \frac{[G(\omega') - G(\omega_\star)]}{\omega' - \omega_\star} \frac{d\omega'}{\omega' - \omega}. \qquad (32)$$

In practice, the reference point is often chosen as $\omega_\star = 0$, thus subtracting the constant $G(0)$. The other possible choice is to take $\omega_\star \to \infty$, then

$$G(\omega) = G(\infty) + \frac{1}{\pi i} \mathcal{P} \int_{-\infty}^{\infty} [G(\omega') - G(\infty)] \frac{d\omega'}{\omega' - \omega}. \qquad (33)$$

Taking the real and imaginary parts of (32), one finds again that Re G and Im G determine each other once the subtraction constant is given.

3.3. *Application: Kramers–Kronig relation*

Let us now discuss the implications of the above results in the context of electromagnetic waves in a dielectric medium. According to Maxwell's theory [8], the applied electric field $\mathbf{E}$ produces the dielectric polarisation $\mathbf{P}$ of the material (its dipole moment per unit volume) and induces the field $\mathbf{D} = \mathbf{E} + 4\pi\mathbf{P}$. For time-dependent fields, the dependence between $\mathbf{D}$ and $\mathbf{E}$ is generally linear but nonlocal in time, and thus can be cast [cf. (16)] as $\mathbf{D}(t) = \mathbf{E}(t) + \int_0^\infty g(\tau)\mathbf{E}(t - \tau)d\tau$ for any causal (and isotropic) medium. The real 'memory' function $g(\tau)$ depends on the medium properties. The corresponding relation for the harmonic components ($\propto e^{-i\omega t}$) reads

$$\mathbf{D}_\omega = \varepsilon(\omega)\mathbf{E}_\omega, \qquad \varepsilon(\omega) = 1 + \int_0^\infty g(\tau)e^{i\omega\tau}d\tau. \qquad (34)$$

This complex valued function $\varepsilon(\omega)$ is called the *dielectric permittivity*. Its ω dependence accounts for the dispersion of waves propagating in the medium, as different frequency components travel at different speeds (determined by Re ε). Since $\varepsilon^*(\omega) = \varepsilon(-\omega)$, one readily finds the symmetry relations

$$\text{Re } \varepsilon(-\omega) = \text{Re } \varepsilon(\omega), \qquad \text{Im } \varepsilon(-\omega) = -\text{Im } \varepsilon(\omega). \qquad (35)$$

Applying now Titchmarsh's theorem to $\varepsilon(\omega) - 1$ and taking into account (35), one arrives at the celebrated *Kramers–Kronig relation*

$$\mathrm{Re}\ \varepsilon(\omega) = 1 + \frac{2}{\pi}\mathcal{P}\int_0^\infty \frac{\omega'\mathrm{Im}\ \varepsilon(\omega')}{\omega'^2 - \omega^2}d\omega' \qquad (36)$$

(and similarly for $\mathrm{Im}\ \varepsilon$). This was the first-known dispersion relation.

It is important to state that $\varepsilon(\omega) \to 1$ as $\omega \to \infty$ according to the universal law, which can be established within the model of Section 2. Indeed, the polarisation at large frequencies is due to the motion of 'light' electrons near 'heavy' (static) atoms, which can be described in the harmonic approximation. For a single electron of charge $-e$, the equation of motion is given by (10) with $f(t) = -e\mathbf{E}_\omega/m$, yielding $\mathbf{r}_\omega = (\omega^2 - \omega_0^2 + i\gamma\omega)^{-1}e\mathbf{E}_\omega/m$ for the displacement vector of stationary oscillations. The induced dipole moment is $\mathbf{P}_\omega = -en\mathbf{r}_\omega$, where n is the number of electrons per unit volume. In view of the relation $\mathbf{P}_\omega = (\mathbf{D}_\omega - \mathbf{E}_\omega)/4\pi$, we finally obtain

$$\varepsilon(\omega) = 1 + \omega_p^2/(\omega_0^2 - \omega^2 - i\gamma\omega), \qquad \omega_p^2 = 4\pi ne^2/m. \qquad (37)$$

The quantity ω_p is called the plasma frequency. One sees that $\mathrm{Re}\ \varepsilon$ and $\mathrm{Im}\ \varepsilon$ [related through (36)] exhibit strong resonance behaviour near ω_0. At large frequencies, $\varepsilon(\omega)$ takes the universal (model-independent) form [8]

$$\varepsilon(\omega) \approx 1 - \omega_p^2/\omega^2 \qquad (\omega \gg \omega_p), \qquad (38)$$

as all the phenomenological parameters ω_0, γ drop out. Note that (38) implies the universal linear response $g(\tau) \approx \omega_p^2\tau$ at small times, $\tau \ll \omega_p^{-1}$.

What about $\omega \to 0$? The model suggests that $\varepsilon(\omega)$ tends to some finite value $\varepsilon(0) > 1$, which is indeed the case for the insulators. However, this changes for a conductor, when electrons can move freely and have no binding energy ($\omega_0 = 0$). Expression (37) then shows a pole $\varepsilon(\omega) \approx i\omega_p^2/\gamma\omega$ at $\omega = 0$. As discussed above, such a singular term must be subtracted in order to derive the dispersion relation in this case [see Exercise (5)].

4. Scattering of classical waves

We now consider the scattering problem, in which the incident wave falls upon a single scatterer, and look at the implications of causality in this case. We make rather general assumptions about the wave-scatterer interaction requiring it to be *linear* (thus superposition principle holds); *short-ranged* (vanishing outside the scatterer); and *energy-conserving* (no energy loss or gain). In a stationary situation, the process can be described in terms of the amplitude of the scattering wave at large distances, which is related to that of the incoming wave through the S matrix.

For simplicity, we consider a real scalar field $\psi(\mathbf{r}, t)$ (e.g., acoustic waves) that satisfies the wave equation in the "free" region outside the spherical scatterer of size a. Also, we restrict ourselves to a spherically-symmetric solution (the so-called s-wave with zero angular momentum). As is well known, such a solution has the form $\psi(r, t) = \varphi(r, t)/r$, where the field φ satisfies the one-dimensional (1D) wave equation on a semiline [cf. (20)]:

$$\frac{\partial^2 \varphi(x, t)}{\partial x^2} - \frac{1}{c^2} \frac{\partial^2 \varphi(x, t)}{\partial t^2} = 0 \qquad (x = r > a). \tag{39}$$

The stationary solutions of (39) can be represented in the complex form $\varphi(k, x, t) = [A(k)e^{-ikx} + B(k)e^{ikx}]e^{-ikct}$ (with wave number $k = \omega/c$) or

$$\varphi(k, x, t) = A(k)[e^{-ikx} - S(k)e^{ikx}]e^{-ikct}. \tag{40}$$

This defines the scattering function, $S(k)$, as the ratio of the outgoing (B) to incoming (A) wave amplitudes. Here, the choice of "$-$" corresponds to requiring $S(k) = 1$ in the absence of a scatterer. (Note that then (40) holds down to $x = 0$ and φ vanishes there, since ψ must be regular.)

4.1. *Symmetry properties of $S(k)$ for real k*

The linearity of the interaction ($x \leq a$) implies that the general solution is a superposition $\varphi(x, t) = \int_{-\infty}^{\infty} \varphi(k, x, t)dk = \varphi_{\text{in}}(x, t) + \varphi_{\text{out}}(x, t)$ of stationary solutions representing the incoming and outgoing waves ($x > a$):

$$\varphi_{\text{in}}(x, t) = \int_{-\infty}^{\infty} A(k)e^{-ik(x+ct)}dk,$$
$$\varphi_{\text{out}}(x, t) = -\int_{-\infty}^{\infty} S(k)A(k)e^{ik(x-ct)}dk. \tag{41}$$

Both $\varphi_{\text{in}}(x, t)$ and $\varphi_{\text{out}}(x, t)$ must be real, hence $A(-k) = A^*(k)$ and thus

$$S(-k) = S^*(k). \tag{42}$$

The energy conservation requires that the total energy flux outside the scatterer, integrated over the time, must vanish. In the 1D case [see

Exercise (7)], this leads to $\int_{-\infty}^{\infty} (\partial\varphi/\partial t)(\partial\varphi/\partial x)\big|_{x>a} dt = 0$ yielding

$$\int_{-\infty}^{\infty} k^2 |A(k)|^2 dk = \int_{-\infty}^{\infty} k^2 |S(k)A(k)|^2 dk. \tag{43}$$

In view of arbitrary (admissible) choice of $A(k)$, we must have

$$|S(k)|^2 = S(k)S^*(k) = 1. \tag{44}$$

This is the *unitarity condition*, an important consequence of the energy conservation. It follows that $S(k) = e^{2i\delta(k)}$, so the only effect of the scattering is to change the phase of the outgoing waves [cf. (25)]. Thus

$$\varphi(k,x,t) = -2iA(k)e^{i\delta(k)}\sin[kx + \delta(k)]e^{-ikct}. \tag{45}$$

With the scattering amplitude $T(k)$ introduced by (26), $S(k) = 1 - iT(k)$, one can single out the 'scattered' part of the solution and write

$$\varphi(k,x,t) = A(k)[(e^{-ikx} - e^{ikx}) + iT(k)e^{ikx}]e^{-ikct}. \tag{46}$$

Here, the first term [$\propto \sin(kx)$] stands for the 'regular' part in the absence of the scatterer, whereas the last term describes the scattering. The unitarity condition leads to the so-called "optical" theorem for the intensity:

$$|T(k)|^2 = -2\mathrm{Im}\,T(k) = 4\sin^2\delta(k). \tag{47}$$

Finally, (42) and (44) together yield the last symmetry property

$$S(-k) = 1/S(k), \tag{48}$$

implying that the phase shift is an odd function of k, $\delta(-k) = -\delta(k)$.

4.2. *Analytic continuation to complex k*

We have so far dealt only with real k, and now derive the general properties of $S(k)$ in the entire complex plane by employing the causality principle.

Causality and analyticity in I_+. Let us build up the incoming wave packet having a sharp front, which reaches the surface at time $t = t_0$. Then $\varphi_{\mathrm{in}}(a,t)$ must vanish at $t < t_0$, where we can take $t_0 = -a/c$, and

$$\varphi_{\mathrm{in}}(a,t) = \int_{-\infty}^{\infty} A(k)e^{-ikc(t-t_0)}dk \qquad (t \geq t_0 = -a/c). \tag{49}$$

By Titchmarsh's theorem, this will be satisfied if and only if $A(k)$ is a causal transform. According to the causality condition, *the outgoing wave, φ_{out},* *cannot appear before the incoming wave, φ_{in}, has reached the scatterer, thus*

$$\varphi_{\text{out}}(a,t) = -\int_{-\infty}^{\infty} S_a(k)A(k)e^{-ikc(t-t_0)}dk \qquad (t \geq t_0), \qquad (50)$$

where $S_a(k) \equiv e^{2ika}S(k)$ and $\varphi_{\text{out}}(a,t) = 0$ at $t < t_0$. Therefore, the function $S_a(k)A(k)$ must also be a causal transform. This is only possible if $S_a(k)$ is regular in I_+. In addition, $S_a(k)$ must be bounded in I_+ (due to condition (ii) of Titchmarsh's theorem). Indeed, $S_a(k)A(k)$ satisfies then the relation (30) at $\text{Im}\,k > 0$, which leads to an estimate

$$|S_a(k)| \leq \frac{1}{2\pi|A(k)|} \int_{-\infty}^{\infty} \frac{|A(k')|}{|k'-k|}dk' \leq 1 \qquad (\text{Im}\,k > 0). \qquad (51)$$

As discussed in Ref. [1], the bound $|S_a(k)| \leq 1$ in I_+ cannot be improved.

We therefore conclude that $S_a(k)$ is holomorphic in I_+. This implies that $S(k) = e^{-2ika}S_a(k)$ has an essential singularity as $|k| \to \infty$ (blowing up exponentially). Thus one cannot derive dispersion relations for $S(k)$ itself, but only for $S_a(k)$. Moreover, due to (51), it requires one subtraction that can be taken at the origin, $S_a(0) = S(0) = 1$. Applying relation (30) to $(S_a(k)-1)/k$, we arrive at the analytic continuation of $S_a(k)$ in I_+

$$S_a(k) = 1 + \frac{k}{2\pi i} \int_{-\infty}^{\infty} \frac{(S_a(k')-1)}{k'(k'-k)}dk' \qquad (\text{Im}\,k > 0) \qquad (52)$$

in terms of its values on the real axis. Taking the complex conjugate and using (42), one finds the extension of the symmetry relation to I_+

$$S^*(k) = S(-k^*). \qquad (53)$$

Analytic continuation to I_-. We define $S(k)$ at any point in I_- by extending the relation (48), derived for real k, to complex values:

$$S(k) = S(k'+iK) = 1/S(-k'-iK) = 1/S(-k) \qquad (K < 0). \qquad (54)$$

Since this function tends to $S(k')$ when $K \to 0$, $S(k)$ is extended as an analytic function in the whole plane, regular except for possible poles in I_- corresponding to the zeros of $S(k)$ in I_+, i.e., as a *meromorphic function*. It also follows from (54) that the relation (53) remains valid in I_-. This leads to the extension of the untarity condition (78) to complex k,

$$S(k)S^*(k^*) = 1. \qquad (55)$$

The relations (53)–(55) provide one-to-one correspondence between values of $S(k)$ in different quadrants of the k-plane:

$$S(-k_0^*) = S_0^*, \qquad S(k_0^*) = (S_0^*)^{-1}, \qquad S(-k_0) = S_0^{-1}, \qquad (56)$$

where $S_0 = S(k_0)$ denotes the value at some point k_0. This yields exactly the same symmetry structure for the pole and zero distribution that we have already seen on the model example of Section 2.4 (for one resonance).

The number of poles is the same as that of zeros and may be either finite or infinite. In the latter case, however, they cannot have an accumulation point at finite k (due to the fact that the zeros of a holomorphic function are isolated points), so their absolute values must approach infinity [1].

4.3. *Canonical product expansion*

A general representation for $S(k)$ in terms of its poles and zeros can be constructed by taking into account the above analytic properties. Let us introduce the so-called *Blaschke product* over all the poles and zeros

$$B(k) = \prod_n B_n(k) = \prod_n \frac{1 - k/k_n^*}{1 - k/k_n}. \qquad (57)$$

Explicitly, the poles k_n and $-k_n^*$ are paired together, so each such a pair gives a contribution (24), whereas each pole $-iK_m$ on the negative imaginary axis yields $(iK_m-k)/(iK_m+k)$. (One can also show that (57) converges if the number of poles is infinite.) By construction, we have $|B(k)| \leq 1$ in I_+ and $|B(k)| \geq 1$ in I_-, i.e., the same bounds as for $S_a(k)$. Then $F(k) = S_a(k)/B(k)$ must be an entire function of k without any zeros; furthermore, $|F(k)| = 1$ on the real axis and $|F(k)| \leq 1$ in I_+. The only possible choice is $F(k) = e^{2ik\epsilon}$, with $\epsilon > 0$. This gives the representation

$$S(k) = e^{-2ik\alpha} B(k) = e^{-2ik\alpha} \prod_n \frac{1 - k/k_n^*}{1 - k/k_n}, \qquad (58)$$

where $\alpha = a-\epsilon \leq a$ is the "effective radius" of the scatterer. (The inequality here is compatible with causality, but in most cases $\alpha = a$ is fulfilled [1].)

4.4. *Causal inequality: Time delay*

It is clear from the above discussion that the poles contain all the information about the internal scatterer. They are located in I_- and admit the physical interpretation as resonances. The product expansion (58) can also be converted into a sum rule, by considering the phase shift derivative

$$\frac{d\delta}{dk} = \frac{1}{2i}\frac{S'(k)}{S(k)} = -\alpha - \sum_n \frac{\mathrm{Im}\, k_n}{|k - k_n|^2} \geq -\alpha \geq -a. \qquad (59)$$

This inequality can be given a clear physical meaning in terms of the time duration associated with the scattering event (Wigner, [1, Section 2.11]).

Let us form an incoming wave packet $\varphi_{\mathrm{in}}(x, t) \equiv \widetilde{\varphi}(x + ct)$ the spectral profile of which, $A(k)$, has a sharp peak around some k_0. As a result of the scattering, the outgoing wave acquires the scattering phase factor $e^{2i\delta(k)}$ under the integral in (41). Applying the stationary phase principle, one can then approximate $\delta(k) \approx \delta(k_0) + (k - k_0)\delta'(k_0)$ and write

$$\varphi_{\mathrm{out}}(x, t) \approx -e^{2i\delta(k_0) + ik_0(x - ct)} \int_{-\infty}^{\infty} A(k)e^{i(k - k_0)[x - ct + 2\delta'(k_0)]}dk. \qquad (60)$$

The integral here can be again reduced to the very function $\widetilde{\varphi}$ (up to a constant phase factor) as $\widetilde{\varphi}(x - ct + 2\delta'(k_0))$. In the absence of the scatterer ($\delta = 0$), we would have $\widetilde{\varphi}(x - ct)$ for the outgoing wave. Thus, the scatterer gives rise to a time delay $\tau_{\mathrm{d}}(k) = 2\delta'(k)/c$ that is expressed through (59).

Clearly, there is no upper bound for $\tau_{\mathrm{d}}(k)$, since the incident wave packet can be "captured" and retained for an arbitrary long time. However, causality does not allow an arbitrary large negative delay (time advance), limiting it by the classical time of flight ($\tau_{\mathrm{d}}(k) > -2a/c$).

5. Scattering of quantum waves

We now carry out a similar program in the case of quantum scattering. For the consistency of presentation, we first introduce an equation for "matter waves" in a short heuristic way, highlighting the particle-wave analogy as well as the differences that arise (see Ref. [9] for the in-depth introduction).

Classically, a particle moving in a conservative potential field $U(\mathbf{r})$ is described by a Hamiltonian function defining the total energy

$$E = \frac{\mathbf{p}^2}{2m} + U(\mathbf{r}), \qquad (61)$$

where $\mathbf{p}$ is the particle momentum. For given E, the motion is restricted to the regions with the positive kinetic energy, $\mathbf{p}^2/2m > 0$. Hence, the condition $E - U(\mathbf{r}) = 0$ sets the so-called turning points separating the classically forbidden region from that with motion. For example, the 1D motion can be either *finite* (bounded between two turning points) or *infinite* (with only one turning point, so that the particle escapes to infinity).

Quantum mechanically, the particle is associated with a wave packet of matter such that its wavelength $\lambda = 2\pi/k$ determines the particle momentum, $\mathbf{p} = \hbar \mathbf{k}$, and its energy, $E = \hbar\omega$, where $\hbar$ is the fundamental physical constant (*Planck's constant*). Let us note the consistency of such a choice; as a whole the packet moves with a group velocity

$$\mathbf{v}_g = \frac{\partial \omega}{\partial \mathbf{k}} = \frac{1}{\hbar}\frac{\partial(\hbar\omega)}{\partial \mathbf{k}} = \frac{\partial E}{\partial \mathbf{p}} = \frac{\mathbf{p}}{m}, \tag{62}$$

which is equal to the particle velocity in the classical sense.

To derive the corresponding "wave" equation for a matter field ψ, we note that for a plane wave $\propto e^{i(kx-\omega t)}$ the values of E and p_x are obtained by acting with the time and coordinate derivatives, respectively. This suggests replacing $E \to i\hbar\partial/\partial t$ and $\mathbf{p} \to -i\hbar\nabla$ in (61), which yields

$$i\hbar\frac{\partial \psi(\mathbf{r},t)}{\partial t} = \left[-\frac{\hbar^2}{2m}\nabla^2 + U(\mathbf{r})\right]\psi(\mathbf{r},t). \tag{63}$$

This is the celebrated *Schrödinger equation*. It is linear (we wanted so), thus the superposition principle holds as for the classical wave equation. However, there are two important differences in the quantum case. Formally, there is a change $i\frac{\partial}{\partial t} \leftrightarrows \frac{\partial^2}{\partial t^2}$, thus the complexness of ψ does matter now (it is not just a "trick", as for classical waves). Conceptually, one defines the quantity $\rho(\mathbf{r},t) = |\psi(\mathbf{r},t)|^2$ as the *probability density* of finding the particle at point $\mathbf{r}$ and time t. Then in agreement with the mass conservation, one can represent (63) in the equivalent form of a continuity equation

$$\partial\rho/\partial t + \operatorname{div}\mathbf{j} = 0, \qquad \mathbf{j} = -i(\hbar/2m)(\psi^*\nabla\psi - \psi\nabla\psi^*), \tag{64}$$

where $\mathbf{j}$ denotes the probability density current.

We have already seen the important role played by stationary solutions in the classical case. In quantum mechanics, a *stationary state* is defined similarly as the solution with a definite energy E, $i\hbar\partial\psi/\partial t = E\psi$. For such states, $\psi(\mathbf{r},t) = \psi_E(\mathbf{r})e^{-iEt/\hbar}$ and the density $|\psi(\mathbf{r},t)|^2 = |\psi(\mathbf{r})|^2$ does not depend on time. Thus, they satisfy the time-independent equation

$$\left[-\frac{\hbar^2}{2m}\nabla^2 + U(\mathbf{r})\right]\psi_E(\mathbf{r}) = E\psi_E(\mathbf{r}). \tag{65}$$

It follows that ψ_E is an eigenfunction of a certain Hermitian operator (*Hamiltonian*) corresponding to a real eigenvalue E. Due to its physical interpretation, the wave function ψ_E must be a single-valued and continuous function everywhere in space. Furthermore, its spatial derivative must also be continuous (unless the potential U has δ-like singularities, see below), and $|\psi_E|^2$ must be bounded. These conditions imply the existence of *only two* types of stationary solutions, which display very different behaviours at infinity. Namely, assuming that $U(\mathbf{r})$ vanishes as $|\mathbf{r}| \to \infty$, one distinguishes between **scattering** states $\propto e^{\pm ik|\mathbf{r}|}$ at *any positive* energy $E = (\hbar^2/2m)k^2$ and **bound** states $\propto e^{-\kappa_n|\mathbf{r}|}$ ($\kappa_n > 0$), which exist only at *certain negative* energy values $E_n = -(\hbar^2/2m)\kappa_n^2$. This gives rise to the continuous and discrete energy spectrum, respectively.

5.1. *Discrete versus continuous spectrum in 1D*

Let us now focus on the 1D case, and consider how the scattering and bound states appear in the problem with a real cutoff potential, $U(x) = 0$ for $|x| \geq a$. Introducing $k = \sqrt{2mE/\hbar^2}$ and $V(x) = 2mU(x)/\hbar^2$, we cast the Schrödinger equation (65) in the relevant regions as follows

$$\psi'' + (k^2 - V)\psi = 0 \quad (|x| < a), \qquad \psi'' + k^2\psi = 0 \quad (|x| \geq a) \tag{66}$$

and look for all solutions which satisfy the conditions specified above [2, 9].

Case $E > 0$, particle incident from the left. In the "free" region ($V = 0$), the general solution is $\psi(k,x) = Ae^{ikx} + Be^{-ikx}$ at $x < -a$, and $\psi(k,x) = Ce^{ikx}$ at $x > a$. The ratios $r = B/A$ and $t = C/A$ determine the amplitudes of the reflected and transmitted wave, respectively. In the "interaction" region ($V \neq 0$), we have $\psi(k,x) = u(k,x)$ with some function u at $|x| < a$.

The constants r, t and the function u can be found in general at *any* $k > 0$ from the *matching conditions* (continuity of ψ and ψ') at the interface:

$$\psi(a - 0) = \psi(a + 0), \qquad \psi'(a - 0) = \psi'(a + 0) \tag{67}$$

(and similarly at $x = -a$). The choice of A depends on normalisation, one usually takes $A = 1$ (the other useful choice is to fix the incoming flux $= 1$).

Since the density $|\psi|^2$ does not depend on time for the stationary states, we have $\partial|\psi|^2/\partial t = 0$ and thus $dj/dx = 0$ by the continuity equation (64). Therefore, one gets the steady probability current, $j(x) = \text{constant}$,

$$j = \frac{\hbar}{2m}(-i\psi^*\psi' + \text{c.c.}) = \frac{\hbar k}{m}|A|^2 \times \begin{cases} 1 - |r|^2, & x < -a \\ |t|^2, & x > a \end{cases}, \tag{68}$$

yielding the important relation $|r|^2 + |t|^2 = 1$ (the flux conservation).

Case $E > 0$, particle incident from the right. Similarly, one can write $\psi(k, x) = e^{-ikx} + r'e^{ikx}$ at $x > a$ and $\psi(k, x) = t'e^{-ikx}$ at $x < -a$ (we put $A = 1$), whereas $\psi(k, x) = \tilde{u}(k, x)$ with some function $\tilde{u}$ at $|x| < a$. This case differs from the previous one by interchanging the incoming and outgoing waves, thus by reversing the time. Since U is real, time reversal for Equation (63) is equivalent to taking the complex conjugate, thus one can relate the coefficients for going from left-to-right and from right-to-left. In particular, the transmission amplitude $t' = t$ [see Exercise (9)].

Case $E < 0$, $k = +i|k|$. In this case, only the exponentially decaying solution $\psi(k, x) = C_+ e^{-|k|x}$ at $x > a$ and $\psi(k, x) = C_- e^{+|k|x}$ at $x < -a$ must be kept to ensure ψ is bounded at infinity. Note that the matching conditions (67) can be equivalently written for the logarithmic derivative

$$-\frac{1}{\psi}\frac{d\psi}{dx}\bigg|_{x=-a} = \frac{1}{\psi}\frac{d\psi}{dx}\bigg|_{x=a} = -|k|. \qquad (69)$$

This is actually a transcendental equation on $|k|$, and therefore may be satisfied only at some *discrete* values $|k_n| = \kappa_n$ (corresponding to the discrete *negative* energies $E_n = -\hbar^2\kappa_n^2/2m$). Clearly, such a bound state has zero current density, $j = 0$. Thus, the bound states carry no flux and correspond, in this sense, to the classical finite motion. The wave function $\psi_n(x)$ is then spatially localised and, therefore, can be normalised according to the usual condition for the probability density, $\int_{-\infty}^{\infty} |\psi_n(x)|^2 dx = 1$.

Example. Consider scattering on a potential $U(x) = -q\delta(x)$, $q > 0$, which can be thought as a limit of a very thin and deep potential well. By integrating (65) around $x = 0$, one finds that $d\psi/dx$ has a jump

$$\psi'|_{x=+0} - \psi'|_{x=-0} = -2\kappa\psi(0), \qquad \kappa = mq/\hbar^2 > 0, \qquad (70)$$

at the singularity. Condition (70) replaces the second one in (67).

For $E > 0$, $\psi(x) = e^{ikx} + re^{-ikx}$ at $x < 0$ and $\psi(x) = te^{ikx}$ at $x > 0$. The matching gives $1 + r = t$ and $ik - ik(1 - r)/(1 + r) = -2\kappa$, yielding

$$r = \frac{i\kappa}{k - i\kappa}, \qquad t = \frac{k}{k - i\kappa}. \qquad (71)$$

For $E < 0$, $\psi(x) = Ce^{-|k||x|}$, and (70) yields $|k| = \kappa$. ($C = 1/\sqrt{\kappa}$ by the normalisation.) Extending (71) to complex k, one sees that these amplitudes have a pole $k = +i\kappa$ located on the *positive* imaginary axis. This illustrates the *new feature* of quantum waves, as compared to classical waves, which is associated with the presence of *bound* states.

The case of a barrier, $U(x) = +q\delta(x)$, $q > 0$, corresponds to (71) under the replacement $\kappa \to -\kappa$. Note that there is no bound state then. Both r, t have a pole $k = -i\kappa$ on the *negative* imaginary axis, the so-called *antibound* (or *virtual*) state. (Note that a formal stationary solution for such a state would diverge exponentially as $|x| \to \infty$.)

5.2. $S(k)$ *and its symmetry properties*

Let us now discuss the analytic properties of the scattering amplitudes in a general setting. To make a connection with the material of Section 4, we consider s-wave scattering, i.e., the 1D problem on a semiline,

$$\psi'' + (k^2 - V)\psi = 0 \quad (0 < x < a), \qquad \psi'' + k^2\psi = 0 \quad (x \geq a). \quad (72)$$

Such a solution is regular at zero, $\psi(0) = 0$ (note that the slope $\psi'(0)$ may be arbitrary), and can be analysed for arbitrary potential V.

Introduce two new solutions $f_{\mp}(k, x)$ of (72) with the behaviour

$$f_{\mp}(k, x) = e^{\mp ikx} \qquad (x > a). \quad (73)$$

These functions are linearly independent, since the Wronskian $W[f_-, f_+] \equiv f_-f_+' - f_-'f_+ = 2ik \neq 0$. Since (72) is invariant under $k \to -k$ or under taking the complex conjugate (V is real), one gets the symmetry relations

$$f_{\mp}(k, x) = f_{\pm}(-k, x) = f_{\pm}^*(k, x). \quad (74)$$

Therefore, the regular solution can be written as a linear combination

$$\psi(k, x) = a(k)[f_-(-k, 0)f_-(k, x) - f_-(k, 0)f_+(k, x)] \quad (x > 0)$$

$$\equiv A(k)[e^{-ikx} - S(k)e^{ikx}] \quad (x \geq a). \quad (75)$$

This yields an important representation for the S function

$$S(k) = \frac{f(k)}{f(-k)}, \qquad f(k) \equiv f_-(k, 0). \quad (76)$$

in terms of the so-called *Jost function* $f(k)$. Making use of the symmetry relations (74), it follows at once that $S(k)$ at real k satisfies

$$S(k)S(-k) = 1, \qquad S^*(k) = S(-k). \quad (77)$$

This leads to the conventional unitarity condition

$$S^*(k)S(k) = 1 \quad \Leftrightarrow \quad S(k) \equiv e^{2i\delta(k)}. \quad (78)$$

These relations are exactly the same symmetry relations (42), (44), and (48) as for the classical waves. Where is the difference?

5.3. *Causality and analytic continuation*

To answer the question, let us turn to the time-dependent problem. For s-waves, the general solution of (63) outside the scatterer ($x > a$) can be again represented as a superposition $\psi(x,t) = \psi_{\text{in}}(x,t) + \psi_{\text{out}}(x,t)$,

$$\psi_{\text{in}}(x,t) = \int_0^\infty A(E)e^{-ikx-iEt}dE,$$

$$\psi_{\text{out}}(x,t) = -\int_0^\infty S(E)A(E)e^{ikx-iEt}dE, \tag{79}$$

of the incoming and outgoing wave packets. (Henceforth, $\hbar = m = 1$.) Though similar to (41), a crucial difference is that now the integrations run only over the positive values, since the energy E is non-negative. This leads to far-reaching consequences in the whole treatment.

Causality condition. In formulating a causality condition for quantum scattering, one needs to take into account that now there is no limiting velocity (due to the nonlinear dispersion $E = k^2/2$) and, as a result, no wave packets can be built which propagate with sharp wave fronts for any finite length of time. (Formally, the "one-sided" ($E > 0$) Fourier integrals cannot vanish over any finite time interval.) This can be incorporated by employing the causality condition in a modified form [1, p.76]: *The scattered wave packet at a distance x at time t cannot depend on the behaviour of the incoming wave packet at x at later times $t' > t$.* Therefore, $\psi_{\text{out}}(a,t)$ at the point $x = a$ must be a linear functional of $\psi_{\text{in}}(a,t)$ such that

$$\psi_{\text{out}}(a,t) = -\int_{-\infty}^\infty g(t-t')\psi_{\text{in}}(a,t')dt', \qquad g(\tau) = 0 \quad (\tau < 0). \tag{80}$$

Let us write $g(\tau)$ as a Fourier integral (note the lower limit of $-\infty$)

$$g(\tau) = \frac{1}{2\pi}\int_{-\infty}^\infty G(E)e^{-iE\tau}dE \tag{81}$$

and also extend the lower integration limit in (79) to $-\infty$ by introducing the step function $\theta(E)$ there. This restores the conventional form of the FT, yielding the relation $\theta(E)S(E)A(E)e^{ika} = 2\pi G(E)\theta(E)A(E)e^{-ika}$ between the FTs of the outgoing and incoming waves. In view of (81), we finally arrive at a causal representation of $S(E)$ at $E > 0$ in terms of $g(\tau)$,

$$S(E)e^{2ika} = \int_0^\infty g(\tau)e^{iE\tau}d\tau. \tag{82}$$

This clearly shows that $S(E)$ is analytic in the upper E half-plane, $I_+(E)$.

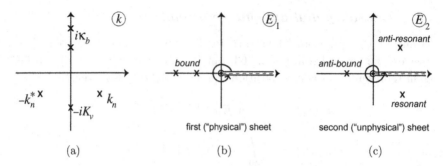

Fig. 1. (a) Schematic representation of the poles in the complex k-plane, picturing two bound states $(i\kappa_b)$, one anti-bound state $(-iK_v)$, and one resonance pair $(k_n, -k_n^*)$. Their corresponding positions on the Riemann surface $k = \sqrt{2E}$ are shown on (b) and (c). The surface has two sheets, each cut along $\operatorname{Re} E > 0$, the lower edge of the slit on the first sheet (b) being joined to the upper edge of the slit on the second sheet (c).

Analytic continuation. Since $k = \sqrt{2E}$, $I_+(E)$ corresponds to the first quadrant of the k-plane, $0 < \arg k < \pi/2$. In contrast to the classical case, causality does not imply anything about the possibility of continuing beyond it, which would correspond to the continuation of $S(E)$ in $I_-(E)$. Moreover, since the mapping from k to E is two-to-one, S is a function of E on a two-sheeted Riemann surface, see Fig. 1. To deal with a single-valued function, we shall consider $S(k)$ which, in the present case of potential scattering, is represented by (76). Note that the functions $S(k)$ and $[S^{-1}(k)]^*$ coincide on the whole real axis, and $S(k)$ is analytic in the first quadrant. Then by Schwarz's reflection principle [7] and due to the derived symmetry properties, the function $S(k) = [S^*(k^*)]^{-1}$ gives the analytic continuation to complex k. Thus, the generalised unitarity condition (55) holds in the whole k-plane. The relations (56) can be used to establish the correspondence between values of S in different quadrants precisely as it was done for the classical field. In particular, $S(k)$ is also regular in the second quadrant. However, $S(k)$ may now have *poles* on the *positive* imaginary k-axis (negative real E-axis), which correspond to *bound* states [cf. (71)].

There is a simple argument supporting the above pole topography [2]. Let $S(k)$ have a pole at some point $k_0 = k_1 + ik_1$. According to (75), the wave function at this point contains only the outgoing wave asymptotically, $\psi(x,t) = u(x)e^{-iEt} \sim e^{i(k_1+ik_2)x - i(k_1^2 - k_2^2 + 2ik_1k_2)t}$. On the other hand, it must satisfy the continuity equation (64). By integrating the latter over the interval $[0, X]$ and choosing X sufficiently large to be in the asymptotic regime, elementary calculations give $k_1 k_2 \int_0^X |u(x)|^2 dx = -k_1 e^{-2k_2 X}$. This condition may be satisfied only if (i) $k_1 = 0$, i.e., the pole of $S(k)$ lies on the imaginary axis; or (ii) $k_1 \neq 0$, $k_2 < 0$, i.e., it lies in the lower half-plane.

The analysis can be made rigorous by considering the analytic properties of the function $f_-(k, x)$ and, therefore, the Jost function that defines the S function through (76). By introducing a suitable Green's function, one can transform the differential equation (66) into an equivalent integral equation that incorporates the boundary condition (73) for $f_-(k, x)$. In the case of a cutoff potential, the Green's function of the "free" equation is given just by (18) (where $\omega, t \to k, x$ and, obviously, $\gamma = 0$), yielding

$$f_-(k, x) = e^{-ikx} + (1/k) \int_x^a \sin[k(x' - x)]V(x')f_-(k, x')dx'. \qquad (83)$$

This type of equations can be solved by iterations, providing us with the analytic continuation of $f(k)$ to complex k [1]. In particular, one can find the asymptotic behaviour of $f(k)$ for large $|k|$ [see Exercise (11)] and establish that $|S_a(k)| = |S(k)e^{2ika}|$ is bounded as $|k| \to \infty$ in I_+.

Product expansion. As compared to the estimate (51) for classical waves, the weaker restriction on $S_a(k)$ in the quantum case is clearly due to the bound states. Their contribution, however, can be again taken into account by means of the Blaschke factors $(k + i\kappa_b)/(k - i\kappa_b)$ ($\kappa_b > 0$). As a result, one can summarise all the above properties in the following canonical product expansion for the S function for quantum scattering [1]:

$$S(k) = e^{-2ika} \prod_n \frac{1 - k/k_n^*}{1 - k/k_n} \prod_{K_v > 0} \frac{1 - k/iK_v}{1 + k/iK_v} \prod_{\kappa_b > 0} \frac{1 + k/i\kappa_b}{1 - k/i\kappa_b}. \qquad (84)$$

It differs from (58) only by (possible) presence of the last product factor.

5.4. $S(E)$ and scattering resonances

The product expansion (84) represents $S(k)$ as a meromorphic function determined fully by its poles. The latter form three different groups: the possible poles on the positive imaginary axis; the ones on the negative imaginary axis; and the complex (symmetrical) poles in I_-.

In the energy domain, $S(E)$ is regular on the first (so-called "physical") sheet except for poles on the negative real axis, corresponding to the energies of bound states. On the second sheet, $S(E)$ can also have poles on the negative real axis, which are then associated with anti-bound states. Bound (as well as anti-bound) states may or may not appear in a specific problem; in any case, their total number can be shown to be finite for a cutoff potential [1]. Generally, such poles may give rise to noticeable energy variations of $S(E)$ only at low scattering energies, $E \approx 0$. Therefore, when $E > 0$ is far from the origin, the overall influence of such poles amounts to a smooth contribution to the scattering phase. The resonance energy behaviour in such a case is entirely due the complex poles of $S(E)$ which arise in complex conjugate pairs (corresponding to k_n and $-k_n^*$, see Fig. 1).

To single out such a resonance contribution, one can keep explicitly only the relevant dominant Blaschke factors in (84) and replace the product over all the other (non-resonant) terms by an overall phase factor, $e^{2i\phi(E)}$. Assuming N resonances contributing at given E, this yields ($k = \sqrt{2E}$)

$$S(E) = e^{2i\phi(E)} \prod_{n=1}^{N} \frac{(k - k_n^*)(k + k_n)}{(k - k_n)(k + k_n^*)} \equiv e^{2i\phi(E)} \prod_{n=1}^{N} \frac{E - \mathcal{E}_n^*(E)}{E - \mathcal{E}_n(E)}. \qquad (85)$$

Here, we have introduced the so-called *complex energy* of nth resonance

$$\mathcal{E}_n(E) = E_n - (i/2)\Gamma_n(E), \qquad (86)$$

where the resonance position, E_n, and its width, Γ_n, are defined as

$$E_n = |k_n|^2/2, \qquad \Gamma_n(E) = -2\sqrt{2E}\mathrm{Im}\, k_n > 0. \qquad (87)$$

Note always negative $\mathrm{Im}\,\mathcal{E}_n$ (thus positive widths) of such complex resonances. Therefore, they can be associated with unstable (decaying) states formed in the internal region ($x < a$) as a result of scattering.

When only one ($N = 1$) narrow resonance is dominant at $E \approx E_n \gg \Gamma_n$, one can approximate (86) by its fixed value $\mathcal{E}_n \equiv \mathcal{E}_n(E_n)$ and also neglect a smooth energy dependence of the background phase ϕ. Then (85) acquires the form $S(E) = e^{2i\phi}[1 - i\Gamma_n/(E - E_n + i\Gamma_n/2)]$, corresponding at $\phi = 0$ exactly to the Breit–Wigner formula [cf. (13), (14), (27)]. At $\phi \neq 0$, the interference between the background and resonance phases gives rise to more general asymmetric resonance profile for the scattering intensity,

$$|T(E)|^2 = \frac{4}{1 + q^2} \frac{(q + \epsilon)^2}{1 + \epsilon^2}, \qquad \epsilon = \frac{E - E_n}{\Gamma_n/2}, \qquad q = -\cot\phi. \qquad (88)$$

This is known as the *Fano formula*, with q being the *shape parameter*. In the limit $|q| \to \infty$, one gets the usual Breit–Wigner profile, whereas the case $q \sim 1$ yields a highly asymmetric shape with a zero at $\epsilon = -q$.

In the general case of many contributing resonances, expression (85) can also be represented as a sum over the separate pole contributions,

$$S(E) = e^{2i\phi}\left[1 - i\sum_{n=1}^{N}\frac{\alpha_n\Gamma_n}{E - \mathcal{E}_n}\right], \qquad \alpha_n \equiv \prod_{m\neq n}\frac{\mathcal{E}_n - \mathcal{E}_m^*}{\mathcal{E}_n - \mathcal{E}_m}. \tag{89}$$

The important difference now is that the factors α_n are generally complex, thus ensuring the unitarity in (89). This gives rise (even at $\phi = 0$) to non-trivial interference between different resonance contributions to the scattering intensity, which cannot be reduced to a simple sum of N Breit–Wigner terms. Such interference effects become especially important when investigating scattering in complex systems with chaotic dynamics (e.g., compound nuclear, semiconductor microstructures, etc.). Such studies require employing statistical methods, which are commonly based on random matrix theory and its applications to open (scattering) systems [10].

5.5. *Dispersion relation with bound states*

Let us now use the general analytic properties of the S function to derive a dispersion relation in the presence of bound states. We have seen above that $S_a(k)$ is analytic everywhere in the upper half-plane I_+, except for a finite number of the poles $i\kappa_b$ on the positive imaginary axis. The latter modify the dispersion relation as compared to the case of classical scattering.

First, we show that *all bound-state poles are simple*, i.e., $\operatorname{res} S(k)|_{i\kappa_b} \neq 0$. According to (76), a bound-state pole $k = i\kappa_b$ of $S(k)$ corresponds to the zero of the Jost function $f(-k)$. Then denoting $\dot{f}(k) \equiv \partial f/\partial k$, one gets

$$\operatorname{res} S(k)|_{i\kappa_b} = -f(i\kappa_b)/\dot{f}(-i\kappa_b). \tag{90}$$

To evaluate $\dot{f}(-i\kappa_b)$, let us take the differential equation (66) for $f_-(k,x)$ and the one obtained by differentiation with respect to k,

$$f_-''(k,x) + (k^2 - V)f_-(k,x) = 0,$$

$$\dot{f}_-''(k,x) + (k^2 - V)\dot{f}_-(k,x) = -2kf_-(k,x).$$

Multiplying the first line by $\dot{f}_-$ and the second one by f_-, and subtracting one from the other, we find that for all k in I_-

$$\int_0^\infty (\dot{f}_- f_-'' - f_- \dot{f}_-'')dx = 2k\int_0^\infty f_-^2(k,x)dx \qquad (\operatorname{Im} k < 0). \tag{91}$$

Here, k can be already extended to I_-, which also ensures the convergence of the integrals at infinity (recall $|f_-| = |e^{-ikx}| = e^{-(\operatorname{Im} k)x}$ at $x > a$). Taking

$k = -i\kappa_b$ and noting that $f(-i\kappa_b) = 0$, the l.h.s. of (91) after the integration by parts becomes $f(k)\dot{f}'(k,0) - \dot{f}(k)f'(k,0) = -\dot{f}(-i\kappa_b)f'(-i\kappa_b,0)$. Now use that the Wronskian $W[f(k), f(-k)] = 2ik$ evaluated at $k = -i\kappa_b$ gives $-f(i\kappa_b)f'(-i\kappa_b,0) = 2\kappa_b$. This leads to

$$\dot{f}(-i\kappa_b) = \frac{2i\kappa_b}{f'(-i\kappa_b)} \int_0^\infty f^2(-i\kappa_b, x)dx = -if(i\kappa_b)\int_0^\infty f^2(-i\kappa_b, x)dx.$$

Substituting this expression into (90), we finally obtain

$$i\,\mathrm{res}\,S(k)\big|_{i\kappa_b} = \left[\int_0^\infty f^2(-i\kappa_b, x)dx\right]^{-1} > 0. \qquad (92)$$

The fact that $f(-i\kappa_b, x)$ is real (hence, the inequality above) is due to the observation that this function a solution of (66) satisfying $f(-i\kappa_b, 0) = 0$ and $f(-i\kappa_b, x) = e^{-\kappa_b x}$ at $x > a$, which are exactly the conditions for the bound-state wave function. The two functions must then coincide up to a constant factor, and $f(-i\kappa_b, x)$ is therefore real. Thus, expression (92) establishes the important relationship between the residues of $S(k)$ at bound-state poles and the normalisation of bound-state wave functions.

We are now ready to derive the dispersion relation for the S function. Since $S_a(k)$ is bounded as $|k| \to \infty$ in I_+, one needs to consider a suitable subtraction. Let us consider the function $F(E) = [S_a(E) - 1]/E$. On the first (physical) Riemann sheet, $F(E)$ is a regular analytic function, except for a finite number N_b of simple poles $\epsilon_n = -\kappa_n^2/2$ on the negative real axis (corresponding to bound states). Moreover, it behaves $F(E) = O(E^{-1})$ as $|E| \to \infty$. Then for any point E taken outside the real axis, we have

$$\int_\gamma \frac{F(E')}{E' - E}dE' = 2\pi i \sum_{n=1}^{N_b} \frac{\mathrm{res}\,S_a(\epsilon_n)}{\epsilon_n(\epsilon_n - E)} + 2\pi i F(E),$$

where the integration contour γ is a circle of a large radius, which encloses all the bound-state poles on the negative real axis and goes continuously along both edges of the cut on the positive real axis. Now, letting the radius tend to infinity and noting that $S(E)$ has a discontinuity across the cut, $S_a(E + i0) - S_a(E - i0) = 2i\,\mathrm{Im}\,S_a(E + i0)$, we arrive finally at the desired relation for the single-valued function $S_a(k)$ valid for all k in I_+:

$$S_a(k) = 1 + \frac{2k^2}{\pi}\int_0^\infty \frac{\mathrm{Im}\,S_a(k')}{k'(k'^2 - k^2)}dk' - 2k^2 \sum_{n=1}^{N_b} \frac{b_n}{\kappa_n^2(k^2 + \kappa_n^2)}, \qquad (93)$$

with $b_n = i\,\mathrm{res}\,S_a(\epsilon_n) > 0$. When k approaches the real axis, one finds again (93), where $S_a(k)$ is then replaced by $\mathrm{Re}\,S_a(k)$ and the integral is taken in the sense of its principle value. This is *Van Kampen's dispersion relation for nonrelativistic particles*. In the absence of bound states, it coincides with the corresponding relation derived for classical waves [cf. (52)].

Finally, we note that the connection between the bound-state poles of $S(k)$ and the eigenvalue problem can be extended to include the resonance states. It requires modifying boundary condition (69) to $\psi'/\psi|_{x=a} = ik$, which simply means that at the resonance pole only the outgoing wave is present away from the scatterer (cf. (22) for its classical analogue). This is the radiative (*Gamow–Siegert*) boundary condition that can be satisfied for certain complex k, which implies that the corresponding operator is no longer Hermitian. Further discussion and explicit constructions of such a non-Hermitian effective Hamiltonian can be found in Refs. [11, 12].

5.6. *Model example: Scattering on a delta-barrier*

As the simplest example illustrating all the essential features of the general problem, we consider *s*-wave scattering on a potential $U(x) = q\delta(x - a)$, where $q > 0$. Since $\psi(0) = 0$, we seek for a stationary solution in the form

$$\psi(x) = \begin{cases} C(k)\sin(kx), & 0 \le x < a, \\ e^{-ikx} - S(k)e^{ikx}, & x > a. \end{cases}$$

The constants C and S are to be determined from the matching conditions (67) and (70) at point $x = a$, which in our case read as follows

$$C(k) = \frac{e^{-ika} - e^{ika}S(k)}{\sin(ka)}, \quad -ik\frac{e^{-ika} + S(k)e^{ika}}{e^{-ika} - S(k)e^{ika}} - k\cot(ka) = 2\frac{mq}{\hbar^2}.$$

Introducing the dimensionless variables $z = ka$ and $\beta = \hbar^2/mqa$, one finds

$$S(z) = e^{-2iz}\frac{2\sin z + \beta ze^{iz}}{2\sin z + \beta ze^{-iz}} \equiv e^{-2iz}\frac{\Delta^*(z)}{\Delta(z)}. \tag{94}$$

Note that such a representation is manifestly unitary (at real z). We now proceed with the analysis (adopted from Refs. [6, 11]) of the obtained result.

When $\beta \to 0$ ($q \to \infty$), we have $S(z) = e^{-2iz}$ and $C(z) = 0$, thus the barrier becomes completely impenetrable for the incident waves. Therefore, the first factor e^{-2iz} in (94) describes the effect of the potential (non-resonant) scattering without the excitation of internal motion. It is the remaining factor, $S_a(z) = e^{2iz}S(z)$, that is entirely responsible for the resonance scattering. Note that the amplitude of the wave function in the internal region is $C(z) = [1 - S_a(z)]e^{-iz}/\sin z$, being thus directly proportional to the resonance part of the scattering amplitude.

The poles of $S(z)$ are given, in view of (94), by the roots of the following complex transcendental equation

$$\Delta(z) \equiv 2\sin z + \beta z e^{-iz} = -ie^{-iz}(e^{2iz} + i\beta z - 1) = 0. \qquad (95)$$

At $\beta = 0$ (impenetrable barrier), we have $\sin z = 0$, i.e., $z_n^{(0)} = \pi n$. These poles correspond to the bounded motion of a free particle trapped in a box.

When $\beta > 0$, there exists a finite probability for the particle to tunnel to the outer (scattering) region. This is reflected in the fact that the real $z_n^{(0)}$ get now converted into complex resonances z_n, with the negative imaginary parts (hence, finite 'lifetimes'). Indeed, all the non-trivial roots of (95) have $\mathrm{Re}\, z_n \neq 0$ and $\mathrm{Im}\, z_n < 0$ (see below). Moreover, since $\Delta^*(z) = 0 = \Delta(-z^*)$ they come in symmetric pairs $(z_n, -z_n^*)$. This yields the representation

$$\Delta(z) = (2+\beta)z \prod_{n=1}^{\infty}\left(1 - \frac{z}{z_n}\right)\left(1 + \frac{z}{z_n^*}\right), \qquad (96)$$

where we have taken into account that $\Delta(z) \approx (2+\beta)z$ as $z \to 0$. When (96) is substituted into (94), $S(z)$ acquires the form (58) of the canonical product expansion, in full agreement with the general discussion.

To analyse the roots of (95), we first note that e^{-iz} is an entire function, which amounts to considering $e^{iz}\Delta(z) = 0$. By writing $z = u + iv$ and separating the real and imaginary parts in this equation, one gets

$$e^{2iz} + i\beta z - 1 = 0 \quad \Leftrightarrow \quad \begin{cases} e^{-2v}\cos 2u = 1 + \beta v \\ e^{-2v}\sin 2u = -\beta u \end{cases}. \qquad (97)$$

Clearly, $v = 0$ (or $u = 0$) implies $u = 0$ (or $v = 0$), therefore (97) has no purely imaginary or purely real non-trivial solutions. Furthermore, it is easy to see that $v = \mathrm{Im}\, z_n < 0$ always at $\beta > 0$. Since such complex roots can be equivalently represented as $z_n = \pi n - (i/2)\ln(1 - i\beta z_n)$, with $n = 1, 2, \ldots$, this suggests to use the following ansatz:

$$z_n = \pi n - \zeta_n \quad (0 \leq \mathrm{Re}\, \zeta_n < \pi), \quad \zeta_n = (i/2)\ln[1 - i\beta(\pi n - \zeta_n)]. \qquad (98)$$

There is one root per each vertical stripe of width π in the complex z-plane.

Further analysis is possible in the limit $\beta \ll 1$ ("high" barrier), corresponding to the case of narrow resonances with small widths (thus long-lived quasistationary states). Making use of the perturbation theory, one can readily find the following limiting behaviour at $\beta\pi n \ll 1$ or $\beta\pi n \gg 1$:

$$z_n \approx \begin{cases} 2\pi n/(2+\beta) - i(\beta\pi n)^2/4, & \beta\pi n \ll 1 \\ \pi(n-1/4) - (i/2)\ln(\beta\pi n), & \beta\pi n \gg 1 \end{cases}. \tag{99}$$

Therefore, the distance $|z_{n+1} - z_n|$ between two consecutive poles approaches π for large n, and their imaginary parts $\operatorname{Im} z_n = O(\log n)$ as $n \to \infty$. This can be shown to hold for an arbitrary cutoff potential [1].

Finally, it is instructive to mention that in the case of $\beta < 0$ (a delta-well) there appears a purely imaginary root $k = i\kappa$ of the corresponding dispersion equation. Furthermore, $\kappa > 0$ only if $|\beta| < 2$, corresponding then to a *bound* state in the well. At $|\beta| > 2$, it becomes an *anti-bound* state with $\kappa < 0$. We leave the details as an exercise to the reader.

6. Exercises

(1) Verify expression (18) for Green's function of the damped oscillator.
(2) Show that any linear combination of causal transforms with constant coefficients is also a causal transform. Verify this for function (17).
(3) For $\operatorname{Im} G(\omega) = \gamma\omega/[(\omega_0^2 - \omega^2)^2 + \gamma^2\omega^2]$ ($\gamma > 0$), find $\operatorname{Re} G(\omega)$.
(4) Show that the Kramers–Kronig relation together with the known universal behaviour $\varepsilon(\omega) \approx 1 - \omega_p^2/\omega^2$ of the dielectric permittivity at large ω imply the following exact sum rule: $\int_0^\infty \omega \operatorname{Im} \varepsilon(\omega)\, d\omega = \pi\omega_p^2/2$.
(5) The causal transform (34) links the complex function $\varepsilon(\omega)$ with the real response $g(\tau)$. In the case of a conductor, the dielectric permittivity $\varepsilon(\omega)$ has a pole at $\omega = 0$: $\varepsilon(\omega) \approx i\alpha/\omega$ ($\alpha > 0$) as $\omega \to 0$.

    (a) Find a relation between the constant α and the limit of $g(\tau)$ as $\tau \to \infty$. Show that the 'regularised' permittivity $\widetilde{\varepsilon}(\omega) = \varepsilon(\omega) - i\alpha/\omega$ satisfies the conventional dispersion relations.

    (b) Compute $\operatorname{Re} \varepsilon(\omega)$, if $\operatorname{Im} \varepsilon(\omega) = \gamma\omega_p^2/[\omega(\omega^2 + \gamma^2)]$ ($\gamma > 0$), and then determine $g(\tau)$ at all times.

(6) $\operatorname{Im} \varepsilon(\omega) = \omega_p^2 \operatorname{Im} G(\omega)$, with $\operatorname{Im} G(\omega)$ from Exercise (3). Show then that $\varepsilon(\omega)$ is given by (37). Find the memory function $g(\tau)$ at all times, and compare the obtained result with that of the previous exercise.
(7) Show that the classical wave equation, $c^2\nabla^2\psi - \partial^2\psi/\partial t^2 = 0$, can be written equivalently as a continuity equation $\partial w/\partial t + \operatorname{div} \mathbf{s} = 0$, with the energy density $w = [c^{-2}(\partial\psi/\partial t)^2 + (\nabla\psi)^2]/2$ and the energy density current $\mathbf{s} = -(\partial\psi/\partial t)\nabla\psi$. Use this to establish (43).

(8) Derive (51) and show that the choice $A(k) = 1/(k - \varsigma)$ (Im $\varsigma < 0$) gives the upper bound $|S_a(k)| \leq 1$ in I_+.

(9) Consider quantum scattering on a cutoff potential (66) in 1D. Show that (at a given energy) the transmission amplitude for a wave incident from the right is the same as for a wave incident from the left. Find further a relation between the corresponding reflection amplitudes.

(10) For $U(x) = U_0\theta(a - |x|)$, $U_0 > 0$, compute the reflection and transmission amplitudes. Show further that in the limit $a \to 0$, $U_0 \to \infty$, but $2aU_0 = q$ kept finite, one reproduces the expressions from (71).

(11) Derive Equation (83). By using a partial integration further, establish the following bounds on the Jost function as $|k| \to \infty$: $|f(k) - 1| \to 0$ (k in I_-) and $f(k) \approx 1 - [V(a-)/4k^2]e^{-2ika}[1 + O(k^{-1})]$ (k in I_+), where $V(a-) \neq 0$ is the discontinuity at the cutoff point.

(12) Find the reflection and transmission amplitudes for a potential $U(x) = q[\delta(x) + \delta(x - a)]$, $q > 0$. Determine the *real* energies for which there exists no reflection from the barrier. Discuss a resonance behaviour both in reflection and transmission, and derive a transcendental equation for the corresponding *complex* resonances.

(13) Show that the wave functions of the 1D discrete spectrum are non-degenerate. Furthermore, for a symmetric potential, $U(-x) = U(x)$, the eigenfunctions $\psi(x)$ can always be chosen to have definite parity, $\psi(-x) = \pm\psi(x)$. (Consider both discrete and continuous spectrum.)

(14) For a double-barrier potential $U(x) = q[\delta(x - a) + \delta(x + a)]$, $q > 0$, construct a *symmetric* solution and find the corresponding reflection (scattering) amplitude. Show that it exhibits a resonance behaviour and derive a transcendental equation for the complex resonances. Assuming further $\beta = \hbar^2/2mqa \ll 1$, determine the real and imaginary parts of the *lowest* resonance to the leading (non-vanishing) order in β.

6.1. *Worked solutions to selected exercises*

(5) (a) $\lim_{\tau \to \infty} g(\tau) = \alpha \neq 0$. Therefore, one can represent $\varepsilon(\omega)$ as follows

$$\varepsilon(\omega) = 1 + \int_0^\infty [\alpha + (g(\tau) - \alpha)]e^{i\omega\tau}d\tau = i\alpha/\omega + \widetilde{\varepsilon}(\omega).$$

By construction, $\widetilde{\varepsilon}(\omega) = 1 + \int_0^\infty (g(\tau) - \alpha)e^{i\omega\tau}d\tau$ is regular everywhere I_+, and satisfies the conditions of Titchmarsh's theorem. It follows then that the Kramers–Kronig relation holds for $\widetilde{\varepsilon}(\omega)$:

$$\operatorname{Re}\widetilde{\varepsilon}(\omega) = 1 + \frac{1}{\pi}\mathcal{P}\int_{-\infty}^{+\infty}\frac{\operatorname{Im}\widetilde{\varepsilon}(\omega')}{\omega' - \omega}d\omega'.$$

(b) $\alpha = \lim_{\omega\to 0}[\omega\operatorname{Im}\varepsilon(\omega)] = \omega_p^2/\gamma$. This gives

$$\operatorname{Im}\widetilde{\varepsilon}(\omega) = \frac{\gamma\omega_p^2}{\omega(\omega^2+\gamma^2)} - \frac{\omega_p^2}{\gamma\omega} = -\frac{\omega_p^2\omega}{\gamma(\omega^2+\gamma^2)}$$

and therefore

$$\operatorname{Re}\varepsilon(\omega) = \operatorname{Re}\widetilde{\varepsilon}(\omega) = 1 - \frac{\omega_p^2}{\pi\gamma}\mathcal{P}\int_{-\infty}^{+\infty}\frac{\omega'\,d\omega'}{(\omega'-\omega)(\omega'^2+\gamma^2)}.$$

The integral here is computed by using Cauchy's residue theorem (the integration contour needs to be closed in I_+), yielding

$$\mathcal{P}\int_{-\infty}^{+\infty}(\ldots)d\omega' = 2\pi i\operatorname{res}\big|_{\omega'=i\gamma} + i\pi\operatorname{res}\big|_{\omega'=\omega} = \frac{\pi\gamma}{\omega^2+\gamma^2}.$$

Collecting everything together, one gets $\varepsilon(\omega) = 1 - \omega_p^2/[\omega(\omega+i\gamma)]$.

The memory function $g(\tau)$ is given as the inverse Fourier transform,

$$g(\tau) = \frac{1}{2\pi}\int_{-\infty}^{\infty}[\varepsilon(\omega)-1]e^{-i\omega\tau}d\omega = -\frac{\omega_p^2}{2\pi}\int_{-\infty}^{\infty}\frac{e^{-i\omega\tau}\,d\omega}{\omega(\omega+i\gamma)}.$$

At $\tau < 0$, we have $g(\tau) \equiv 0$ automatically (the contour is closed in I_+). At $\tau > 0$, the contour must be closed in I_-, and computing the sum of two residues at 0 and $-i\gamma$ yields

$$g(\tau) = -\frac{\omega_p^2}{2\pi}\big(-2\pi i\operatorname{res}\big|_{\omega=0} - 2\pi i\operatorname{res}\big|_{\omega=-i\gamma}\big) = \frac{\omega_p^2}{\gamma}(1-e^{-\gamma\tau}) \quad (\tau > 0).$$

This function has a finite limit $g(\tau) \to \omega_p^2/\gamma$ as $\tau \to \infty$, corresponding to a pole $\varepsilon(\omega) \approx i\omega_p^2/\gamma\omega$ at $\omega = 0$. This shows that the medium is a conductor. For an insulator, $\varepsilon(\omega)$ is regular at $\omega = 0$, thus the corresponding memory function must go to zero at large times [e.g., see Exercise (6)]. However, the small τ behaviour is universal, $g(\tau) \approx \omega_p^2\tau$, in both cases.

(9) Let $\psi_L(x)$ and $\psi_R(x)$ be the solutions of the Schrödinger equation for a wave incident from the left and for a wave incident from the right, respectively. Their asymptotic behaviour can be expressed in terms of the corresponding reflection (r and r') and transmission (t and t') amplitudes,

$$\psi_L(x) = \begin{cases} e^{ikx} + re^{-ikx}, & x < -a \\ te^{ikx}, & x > a \end{cases}, \quad \psi_R(x) = \begin{cases} t'e^{-ikx}, & x < -a \\ e^{-ikx} + r'e^{ikx}, & x > a \end{cases}.$$

Since $\psi_L(x)$ and $\psi_R(x)$ satisfy the same equation, their Wronskian must be constant, $W[\psi_L(x), \psi_R(x)] = \psi_L(x)\psi_R'(x) - \psi_L'(x)\psi_R(x) = c$. Evaluating the constant c at $x < -a$ and at $x > a$, one finds $2ikt' = 2ikt$, thus $t' = t$.

Now we can also use that the Schrödinger equation is real, so if $\psi(x)$ is a solution then $\psi^*(x)$ is also a solution for the same energy but with a different asymptotic behaviour. In particular,

$$\psi_L^*(x) = \begin{cases} e^{-ikx} + r^*e^{ikx}, & x < -a \\ t^*e^{-ikx}, & x > a \end{cases}.$$

Consider yet another solution $\tilde{\psi}(x) = \psi_L^*(x) - r^*\psi_L(x)$. It has the following asymptotics:

$$\tilde{\psi}(x) = \begin{cases} (1 - |r|^2)e^{-ikx}, & x < -a \\ t^*e^{-ikx} - r^*te^{ikx}, & x > a \end{cases}.$$

The current conservation implies the relation $1 - |r|^2 = |t|^2$, hence

$$\tilde{\psi}(x) = t^* \begin{cases} te^{-ikx}, & x < -a \\ e^{-ikx} - (r^*t/t^*)e^{ikx}, & x > a \end{cases}.$$

This coincides (up to the overall normalisation) with the asymptotic behaviour for $\psi_R(x)$, thus $t' = t$ (as before) and $r' = -r^*t/t^*$.

References

[1] H. M. Nussenzveig, *Causality and Dispersion Relations*. Academic Press, New York, 1972. The full-text electronic edition is also available at: http://www.sciencedirect.com/science/bookseries/00765392/95.
[2] A. M. Perelomov and Y. B. Zeldovich, *Quantum Mechanics — Selected Topics*. World Scientific, Singapore, 1998.
[3] J. R. Taylor, *Scattering Theory*. John Wiley & Sons, New York, 1972.
[4] M. Reed and B. Simon, *Methods of Modern Mathemematical Physics, vol. III, Scattering Theory*. Academic Press, New York, 1979.
[5] D. Yafaev, *Scattering Theory*, Springer, New York, 2000.
[6] V. V. Sokolov, *Decays and Resonance Phenomena in Classical and Quantum Mechanics*. NSTU, Novosibirsk, 1995 [in Russian].
[7] M. J. Ablowitz and A. S. Fokas, *Complex Variables: Introduction and Applications*. Cambridge University Press, Cambridge, 2003.
[8] J. D. Jackson, *Classical Electrodynamics*. John Wiley & Sons, New York, 1999.
[9] E. Merzbacher, *Quantum Mechanics*. John Wiley & Sons, New York, 1998.

[10] Y. V. Fyodorov and D. V. Savin, *Resonance scattering of waves in chaotic systems*. In eds. G. Akemann, J. Baik and P. di Francesco, *The Oxford Handbook of Random Matrix Theory*, pp. 703–722. Oxford University Press, Oxford, 2011.

[11] D. V. Savin, V. V. Sokolov and H.-J. Sommers, Is the concept of the non-Hermitian effective Hamiltonian relevant in the case of potential scattering? *Phys. Rev. E,* **67** (2003), 026215.

[12] T. J. Christiansen and M. Zworski, A mathematical formulation of the Mahaux–Weidenmüller formula for the scattering matrix, *J. Phys. A: Math. Theor.,* **42** (2009), 415202.

Chapter 3

Modelling — What is it Good For?

Oliver S. Kerr

Department of Mathematics, City University London,
Northampton Square, London EC1V 0HB, UK
o.s.kerr@city.ac.uk

Modelling is at the heart of applied mathematics. Here, we look at some of the basic ideas and methods that underlie many mathematical models such as conservation laws and approximation. Such approximations should be rational and can lead to simpler systems of equations that can be solved more readily. Simplifications can lead to loss of accuracy, but can also give insight into the underlying phenomena. We look at a variety of classic models, some of which have made a significant impact beyond their initial areas of application.

1. Introduction

Mathematical modelling is at the very core of applied mathematics — the development of a set of equations whose solutions represent the behaviour of some real phenomenon. Early models range from the not-very-accurate model of rabbit populations by Leonardo Pisano, better known as Fibonacci, where the number of pairs of breeding rabbits each month satisfies the simple recurrence relation

$$F_n = F_{n-1} + F_{n-2}, \tag{1}$$

to the far more exact laws of motion laid down by Sir Isaac Newton in the 17th century. The relation (1) and its resulting numerical sequence 1, 1, 2, 3, 5, 8, 13, ... has been the subject of much study. This sequence has subsequently been found in nature in unexpected areas such as the number of seed spirals in a sunflower. Newton's laws of motion, on the other hand, have proved to be an exceptionally accurate model for how objects move, and only start to falter in the more extreme circumstances where Einstein's relativity has to be taken into account.

Some modelling, such as that used for weather forecasting, involve a huge amount of work. It is not possible just to use the basic physics of the problem and put it all into a computer as the detail required would be impossible — how would you take into account the interaction of wind over forests? It is not practical to calculate the flow of air through the branches and leaves of each tree. Tackling problems such as the weather firstly requires the development of mathematical models for all the relevant phenomena that may involve a significant amount of simplification. Secondly you have to work out how to solve the resulting equations in a practical way and lastly you have to do the calculations. For massive problems such as the weather, this last part can be a significant hurdle in itself as the computing resources required may be huge. Some problems can now be addressed by solving the complete set of governing equations (which usually involve some standard modelling assumptions) on computers. This removes the need for models that simplify the problem, and perhaps provide an insight into underlying processes, but may demand the organisation and analysis of large amounts of data — this is not the concern of this chapter.

This chapter does not seek to give a self-contained course on mathematical models and modelling. There are many (much longer) text books around which do this for students of different levels [1–3]. Instead, we will focus on other aspects of models such as what you should look for in a model: What can you get out of a model? Does it tell you anything about the problem under consideration? For example, the Fibonacci sequence is not going to be an accurate representation of any rabbit population. All it does tell us is that you may expect rabbit populations to grow roughly exponentially provided there are no restrictions on population levels.

Here, we will focus on a personal choice of several classic models. These models involve a reasonably high level of simplification. This was necessary at the time in order to make them computationally practical, but this does also mean that even though there may be a significant loss of numerical accuracy they do, in general, identify the essence of the problems under consideration. Other models could have been chosen, and there are probably more appropriate choices that could have been made. However, some of these models have shed light on some important aspects of the problems under consideration and have often had far-reaching impacts in mathematics and elsewhere far beyond the initial problem under consideration.

2.   Models in the media

Before concentrating on what models should be, it may be instructive to
see what they should not be. At regular intervals stories appear in the
news media about scientists working out some formula to explain some
phenomenon. A typical example would be the case of the 2005 investigation
of "Beer Goggles" by Nathan Efron, Professor of Clinical Optometry at The
University of Manchester, and later at Queensland University of Technology
in Australia. The following is an extract from the BBC News web site[a]:

> Scientists believe they have worked out a formula to calculate how "beer
> goggles" affect a drinker's vision.
>
> The drink-fueled phenomenon is said to transform supposedly "ugly"
> people into beauties — until the morning after.
>
> Researchers at Manchester University say while beauty is in the eye of
> the beer-holder, the amount of alcohol consumed is not the only factor.
>
> Additional factors include the level of light in the pub or club, the
> drinker's own eyesight and the room's smokiness.
>
> The distance between two people is also a factor.
>
> They all add up to make the aesthetically-challenged more attractive,
> according to the formula.

$$\beta = \frac{(A_n)^2 \times \delta(S+1)}{\sqrt{L} \times (V_0)^2},$$

- A_n = number of units of alcohol consumed,
- S = smokiness of the room (graded from 0–10, where 0 clear air; 10
  extremely smoky),
- L = luminance of 'person of interest' (candelas per square metre; typically 1 pitch black; 150 as seen in normal room lighting),
- V_0 = Snellen visual acuity (6/6 normal; 6/12 just meets driving standard),
- δ = distance from 'person of interest' (metres; 0.5 to 3 metres).

The formula can work out a final score, ranging from less than one — where
there is no beer goggle effect — to more than 100.

This "model" attracted quite a bit of attention in the press at the time.
There are quite a few things that seem to be wrong with this formula,[b] for
example

---

[a]http://news.bbc.co.uk/1/hi/england/manchester/4468884.stm.

[b]This model came before the ban on smoking in pubs and clubs introduced in 2007.

- It is dimensionally inconsistent.
- It does not seem to based on any rational model.
- It does not seem to based on any quantitative observations.
- There is no indication what a score of, say, 5 means, apart from being more than 1 and less than 100.

Unfortunately, there can be a public perception that this is what applied mathematicians do: dream up formulae and apply them arbitrarily to some phenomenon. When developing a model one would hope that it is based on the problem under consideration in a rational way. There are times where an element of guess work may be involved when the underlying processes are unclear, but it would be hoped that at all times there is a clarity to the underlying assumptions that have been made so that others can make a considered evaluation of the nature of your model. However, by its very nature, when modelling you will often be presenting your work to those less mathematically skilled that you are. Just as some will see the formula for "beer goggles" and accept it unquestioningly, so some will accept more rigorous modelling in the same way. You have a duty not to exploit this credulity, but to try and make you work and the underlying assumptions as accessible as possible to the non-specialist.

As an aside, one of the Ig Nobel Prizes of 2013 was awarded to Laurent Bégue, Brad J. Bushman, Oulmann Zerhouni, Baptiste Subra, and Medhi Ourabahthe for their peer-reviewed paper "'Beauty is in the eye of the beer holder': People who think they are drunk also think they are attractive" [4].

3. Fundamentals

At its most simple level a model can be thought of as a mathematical depiction of some phenomenon. We would like this model to be based on what is going on, and it should tell us something about the phenomenon — quantitative or qualitative. All models include some simplification or approximation and we would like this development to be, in some sense, rational.

When attempting to model anything, there are some basic things you will require in order to come to a successful conclusion:

- You need to know what is going on.
- You need to know what physical/governing laws are relevant.
- You need to know what approximations are appropriate and acceptable.
- You need to know how to solve the equations you come up with.

When you first encounter a problem there is no reason why, as a mathematician, you should know all of these things, or possibly even *any* of these things! It is always a good idea to read up about the subject and, if possible, talk to experts in the field. This will help you avoid repeating work previously done by others.

Sometimes the governing laws will be known to be very accurate, in which case you may be able to create a model that is a very good representation of what is going on. Sometimes you will make simplifications that will, by necessity, reduce the accuracy of your results but it is to be hoped will still provide a representation of what is going on, and provide an insight into the phenomena of interest. If you can simplify things with minimal impact on the results then this gives a good indication what is and is not important in the problem. The best models may provide the greatest simplification, but still give a very good representation of reality.

Occasionally you may encounter models that have results that resemble the phenomena of interest, but are not necessarily based on the known processes involved. This is the weakest sort of model. It may be hoped that there may be some link between the model and reality which is yet to be identified, and could lead to further investigations and a better understanding of the problem.

Although covered to a greater depth in text books, we will briefly look at some of the fundamental tools that are common to many examples of modelling. These are the uses of the continuum hypothesis, dimensional analysis and conservation laws.

3.1. *Continuum hypothesis*

For the continuum hypothesis one takes an intrinsically discrete system, and assumes one can describe it in terms of a continuous variable. For example fluids and solids are considered to be continuous media, and not a collection of discrete atoms, or populations are treated as real numbers so their evolution can be described by, say, differential equations, and not as individuals. These assumptions are usually uncontroversial, but care must be taken whenever the resulting model predicts something that makes the continuum hypothesis untenable. An example of this can be found in the analysis of cusp-like structures forming on the surface of viscous fluids investigated by Jeong and Moffatt [5]. They found that in certain circumstances their analysis predicted a radius of curvature of the fluid surface of order 10^{-42} m. As they observed, this is far smaller than the inter-atomic spacing of the fluid and so the continuum hypothesis breaks down. Indeed it is far smaller than the Plank length scale, which in some models of the Universe is the smallest scale where space can be assumed to be continuous! We shall return to the continuum hypothesis in a later section.

3.2. *Non-dimensional numbers and Buckingham's theorem*

Often in problems you will have several parameters that describe the properties of the system. It is usually useful to try and reduce the number of parameters involved. This can often be achieved using Buckingham's theorem [6].

Physical quantities are measured with respect to some reference quantity. Usually these are based on the seven basic SI units[c]:

- Length — metre (m).
- Time — second (s).
- Mass — kilogram (kg).
- Electric current — ampere (A).
- Temperature — kelvin (K).
- Luminous intensity — candela (cd).
- Amount of a substance — mole (mol).

and the many (30+) derived units for other quantities (some of which have names). For example

- Velocity $(m\,s^{-1})$.
- Acceleration $(m\,s^{-2})$.
- Force — newton $(N = kg\,m\,s^{-2})$.
- Pressure — pascal $(Pa = kg\,m^{-1}\,s^{-2})$.
- Energy — joule $(J = kg\,m^2\,s^{-2})$.
- Power — watt $(W = J/s = kg\,m^2\,s^{-3})$.

The basic idea is that when you want to model a system you first draw up a list of the quantities that specify the system: dimensions, speeds, masses, physical properties, and so on. We can then write our problem as

---

[c]There are times when other units may be better as people will find them more accessible. For example, what is your age in seconds?

a mathematical equation

$$f(q_1, q_2, q_3, \ldots, q_n) = 0, \tag{2}$$

where the n quantities are these physical parameters (nothing is being said about the form of f here).

If these n quantities are expressed in terms of k basic physical units then we can re-express our problem as

$$F(\pi_1, \pi_2, \pi_3, \ldots, \pi_p) = 0, \tag{3}$$

where $p = n - k$ and the quantities π_i are dimensionless.

For example, if we wanted to know the drag on a sphere moving through a fluid, then we may list the relevant quantities as the drag force, D, the speed, U, the radius, r, the density of the fluid, ρ, and the dynamic viscosity, μ. These can be described in term of the basic physical quantities length, L, mass, M and time, T, so $k = 3$. Hence, $p = 5 - 3 = 2$ and we can describe our system with two parameters, π_1 and π_2. The parameters, π_i, will be of the form

$$\pi_i = q_1^{m_{1,i}} q_2^{m_{2,i}} \ldots q_n^{m_{n,i}}. \tag{4}$$

The p vectors $(m_{1,i}, m_{2,i}, \ldots, m_{n,i})$ will be linearly independent.

For the sphere, if we use D, U, r, ρ and μ, then we have for each i

$$\pi_i = D^{m_{1,i}} U^{m_{2,i}} r^{m_{3,i}} \rho^{m_{4,i}} \mu^{m_{5,i}}. \tag{5}$$

This has dimensions

$$(MLT^{-2})^{m_{1,i}} (LT^{-1})^{m_{2,i}} (L)^{m_{3,i}} (ML^{-3})^{m_{4,i}} (ML^{-1}T^{-1})^{m_{5,i}}$$

$$= M^{m_{1,i}+m_{4,i}+m_{5,i}} L^{m_{1,i}+m_{2,i}+m_{3,i}-3m_{4,i}-m_{5,i}} T^{-2m_{1,i}1-m_{2,i}-m_{5,i}}. \tag{6}$$

For this to be dimensionless, we require

$$m_{1,i} + m_{4,i} + m_{5,i} = 0, \qquad m_{1,i} + m_{2,i} + m_{3,i} - 3m_{4,i} - m_{5,i} = 0,$$

$$-2m_{1,i}1 - m_{2,i} - m_{5,i} = 0. \tag{7}$$

There are an infinite number of solutions to these equations. Two independent solutions are

$$m_{1,1} = 1, \quad m_{2,1} = -2, \quad m_{3,1} = -2, \quad m_{4,1} = -1, \quad m_{5,1} = 0 \tag{8}$$

and

$$m_{1,2} = 0, \quad m_{2,2} = 1, \quad m_{3,2} = 1, \quad m_{4,2} = 1, \quad m_{5,2} = -1 \tag{9}$$

and the corresponding two non-dimensional parameters are

$$\pi_1 = \frac{D}{\rho r^2 U^2}, \qquad \pi_2 = \text{Re} = \frac{\rho U r}{\mu}. \tag{10}$$

where Re is the well-known Reynolds number of the flow.

In general, our problem can be expressed as a solution to

$$F(\pi_1, \pi_2) = 0. \tag{11}$$

This can be re-arranged as

$$\pi_1 = G(\pi_2) \tag{12}$$

for some function $G(\pi_2)$. This in turn could be re-expressed as the better known expression for drag on a body:

$$D = \rho r^2 U^2 G(\text{Re}). \tag{13}$$

This solution drops out in this analysis because of the conventional choice for π_1 and π_2. Many other possible choices do exist, but it is wise to stick to ones that are familiar to those who know the field. Convention is a powerful force not to be trifled with. The literature or experts in the field will provide guidance if necessary.

3.3. *Conservation laws*

Many of the governing equations of systems are based around conservation laws, for example the rate of change of some quantity of something in some volume is balanced by the net rate of flow of that quantity into the volume combined with its possible creation or destruction.

There are conservation laws that are familiar from physics: the conservation of energy, mass and momentum. These can give rise to well-known governing equations, for example the heat equation, and the continuity and Navier–Stokes equations from fluid mechanics. These, and many other standard equations from physics and engineering, do not need to be derived every time you start a new problem. Many other problems have underlying conservation laws. For example, if you look at traffic on a motorway between junctions then the number of cars obeys a conservation law. The number of cars in the section increases as cars drive in, and decrease as cars drive out. Hopefully, there is no destruction of cars!

In the following sections, we will look at some models that use the continuum hypothesis, Buckingham's theorem and conservation laws to develop models of real problems.

4. Dimensional analysis and the atomic bomb

Dimensional analysis and Buckingham's theorem can be extremely powerful tools in modelling. They can be used to reduce the apparent size of the parameter space of problems, and can enable results obtained from apparently very different circumstance to be applied in unexpected places. An example of this is the analysis of the first atomic explosion by G. I. Taylor [7–9].

A short time after the detonation of the atomic bomb, the fireball is essentially spherical, as shown in the photograph of the explosion in Fig. 1. If you wanted to analyse what such an explosion looked like in this initial phase then you would first list the physical quantities that may be relevant and their dimensions. Such a list could look like

- Energy release $E \sim ML^2T^{-2}$,
- Time since explosion $t \sim T$,
- Radius of the fireball $r \sim L$,
- Density of air at rest $\rho \sim ML^{-3}$,
- Air pressure $p_0 \sim ML^{-1}T^{-2}$, (and other atmospheric properties?)
- Acceleration due to gravity $g \sim LT^{-2}$.

Fig. 1. High speed photograph of the fireball of the Trinity atomic bomb test shortly after detonation on July 16, 1945.

Here, we have six parameters and three basic physical quantities involved, so the process can be reduced to three non-dimensional parameters. However, a bit of thought allows us to simplify matters a bit more. For example, a short time after the blast one might expect that the effect of gravity would be negligible. If this is the case we can neglect g. In the initial stages the blast pressure generated by the bomb would be thousands of times higher than ambient air pressure, with the shock waves moving much faster than speed of sound. This is clearly seen in Fig. 1 where the fireball has expanded by of order 100 m in around 1/60 seconds, far faster than the normal speed of sound in air. If we ignore ambient air pressure, and other atmospheric properties, we have effectively reduced the list to four physical quantities. Hence, an equation for the radius of the fireball can be written as a solution to the equation

$$f(r, E, t, \rho) = 0 \tag{14}$$

for some unknown function f. From Buckingham's theorem we can write (14) as

$$F(\pi) = 0, \tag{15}$$

where

$$\pi = r^a E^b t^c \rho^d. \tag{16}$$

This has dimensions

$$L^{a+2b-3d} T^{-2b+c} M^{b+d}. \tag{17}$$

For this to be dimensionless, we require

$$a + 2b - 3d = 0, \qquad -2b + c = 0, \qquad b + d = 0. \tag{18}$$

Setting (arbitrarily) $b = 1$ we get the non-dimensional group

$$\pi = \frac{Et^2}{\rho r^5}. \tag{19}$$

From much smaller experiments with conventional explosives, G. I. Taylor knew that the zero of $F(\pi)$ is found when π is around one. In other words

$$E \approx \frac{\rho r^5}{t^2}. \tag{20}$$

From Fig. 1, we can estimate

$$E \approx \frac{\rho r^5}{t^2} \approx \frac{(1\,\text{kg}\,\text{m}^{-3})(100\,\text{m})^5}{(0.016\,\text{s})^2} \approx 4 \times 10^{13}\,\text{J}. \tag{21}$$

This is roughly the energy released by 10,000 tons of TNT. This was meant to be top-secret classified information at the time!

5. The continuum hypothesis, conservation laws and traffic

The use of the continuum approximation and conservation laws are involved in the modelling of traffic by Lighthill and Whitham [10]. This makes use of the continuum hypothesis and of a conservation law in a problem where it may not be initially clear that these are appropriate, and indeed may not be in some circumstances.

If you want to know how traffic might behave then you may look at the average density of cars along the road. This will depend on position and time, say $\rho(x, t)$ cars per kilometre. If one considers a portion of the road of length Δx (see Fig. 2) then if the average flow of cars at any point is $q(x, t)$ then we would have a conservation law that the rate of change of cars in this section of road will be given by the net flow of cars into this section:

$$\frac{d}{dt} \int_x^{x+\Delta x} \rho(x, t)\, dx = q(x, t) - q(x + \Delta x, t). \tag{22}$$

Taking the limit $\Delta x \to 0$ gives

$$\frac{\partial \rho}{\partial t} = -\frac{\partial q}{\partial x}. \tag{23}$$

This will be an appropriate approximation where we have length and time scales where ρ and q are reasonably smoothly varying.

To complete the model one needs to relate the traffic flow to the car density. There is no obvious law connecting the two, but some thought gives us a plausible relation we can try. When the density of traffic is low, the cars will move along at some maximum speed, $v_{\max}$, while as the traffic density reaches a maximum, $\rho_{\max}$, the traffic will slow down to nothing. We can assume that $v(\rho)$ will be a continuous decreasing function. This can be achieved simply by using a function of the form

$$v = v_{\max} \left(1 - \left(\frac{\rho}{\rho_{\max}} \right)^\alpha \right). \tag{24}$$

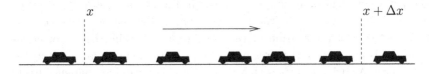

Fig. 2. Traffic moving through a section of road.

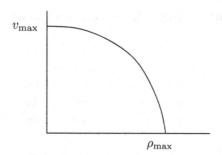

Fig. 3.   Proposed relation of traffic speed to traffic density for $1 \leq \alpha$.

We can apply the restriction $\alpha \geq 1$ to ensure that there is not a sharp drop in traffic speed for low traffic densities. This gives a monotonically decreasing dependency of the traffic speed on the traffic density as required, as illustrated in Fig. 3. The simplest case would be to set $\alpha = 1$, but a more general value is retained with little added complexity.

The flux of cars along the road will be given by

$$q = \rho v = v_{\max}\rho \left(1 - \left(\frac{\rho}{\rho_{\max}}\right)^{\alpha}\right), \tag{25}$$

so

$$\frac{\partial \rho}{\partial t} = -\frac{\partial}{\partial x}\left(v_{\max}\rho\left(1 - \left(\frac{\rho}{\rho_{\max}}\right)^{\alpha}\right)\right) \tag{26}$$

or

$$\frac{\partial \rho}{\partial t} + v_{\max}\left(1 - (\alpha + 1)\left(\frac{\rho}{\rho_{\max}}\right)^{\alpha}\right)\frac{\partial \rho}{\partial x} = 0. \tag{27}$$

On characteristics given by

$$\frac{dx}{dt} = v_{\max}\left(1 - (\alpha + 1)\left(\frac{\rho}{\rho_{\max}}\right)^{\alpha}\right) \tag{28}$$

we observe ρ, and hence also v, will be constant. Thus the slope of the characteristics will be constant as shown in Fig. 4. In general, these characteristics may cross each other. The crossing of the characteristics can be managed by the application of the appropriate Rankine–Hugoniot equations. This can give rise to regions of high traffic density ahead of a discontinuity. As drivers cross such a shock wave they would experience freely moving traffic suddenly giving way to slowly moving densely packed traffic,

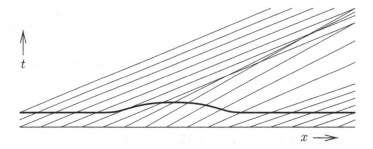

Fig. 4. Characteristics generated by areas of higher traffic density. Note, these characteristics may cross in regions where their gradients increase as x increases.

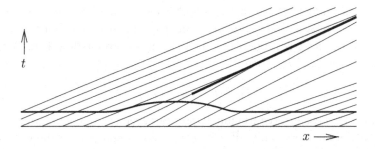

Fig. 5. Shocks forming at the back of areas of higher traffic density.

which then becomes more freely moving for no apparent reason — a puzzling experience that will be familiar to those that drive on motorways at busy times.

It should be noted that the slope of the characteristics can point backwards for high traffic densities and appropriate values of α. Thus, the blocks of slow moving traffic can move in the opposite direction to the direction of motion of the cars. This has been observed in reality.

There are many possible improvements to this model, many of which have been applied to extensions of this problem over the years:

- Speed — use a density/speed relation based on observation.
- Shocks are not sharp — maybe introduce some sort of diffusion?
- Dynamics of speed changes — allow for acceleration.
- Multiple lanes on motorways — do overtaking lanes have an effect?
- Visibility — how far ahead can you see? Does this affect how you drive?

It should be remembered that this model is based on the use of the continuum hypothesis for cars. It is assumed that we are dealing with length and time scales over which the average density of cars will vary relatively smoothly. Thus there is an intrinsic contradiction in its prediction of shock waves. However, over larger scales the behaviour predicted by this model is plausible, and has been observed in real traffic.

6.　Populations and the logistic equation

Here, we look at a simple model that had a profound and wide impact on modelling and how people understood the various ways systems could behave.

A basic model of the population of some animal that has a seasonal breeding pattern is that the increase in the population from one breeding season to the next is proportional to the original population. If the population in cycle n is P_n, then

$$P_n = P_{n-1} + AP_{n-1}. \tag{29}$$

Here, we have used the continuum hypothesis even though populations tend to be whole numbers! This will be a reasonable assumption if the potential population under consideration is large enough.

If A were a positive constant then (29) would give exponential growth of the population. In reality A would depend on many factors such as food availability, predation, overcrowding, age profile (included to some extent in the Fibonacci model), weather and so on, and so it may depend on n, P_{n-1}, and even P_{n-2} and so on. Some of these can be taken into account at the simplest level by making the value of A related to the population level P_{n-1}, with its value decreasing to zero as the population increases to the maximum sustainable level. If we use the simplest decreasing function $A(P) = B - CP$, with B and C positive constants, then the recurrence relation can be re-scaled to give the discrete logistic equation

$$x_n = f(x_{n-1}) = ax_{n-1}(1 - x_{n-1}), \tag{30}$$

with the restriction that $1 \leq a \leq 4$. The properties of this recurrence relation have been much studied and some of the rich phenomena are detailed in Robert May's important paper from 1976 [11]. These include the existence of solutions that consist of steady sequences, periodic sequences and chaotic sequences, and the existence of period-doubling bifurcations as illustrated

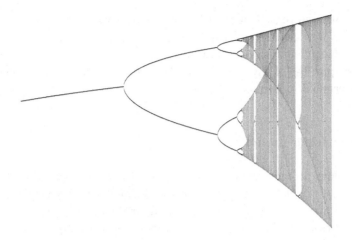

Fig. 6.   Solutions of the logistic equation for $2.5 \leq a \leq 4$.

in Fig. 6. There are also other results such as once period 3 solutions exist there must also co-exist solutions of all other periods.

How applicable are all the rich behaviours to a population modelled by the original relation (29)? The answer is almost certainly "not very". However, this does not mean that this study has been pointless. It can be shown that many of these results are generic to recurrence relations like (30) where $f(x)$ is a one-humped function. Hence, the phenomena observed in the discrete logistic equation are quite general. The addition of random environmental factors meant that any delicate phenomena that, for example, depend on small ranges of a or which take a long time to emerge would almost certainly not be observed in a real population. But more robust behaviours that occur for a larger range of parameters, such as chaotic behaviour are likely to be observable. It is this conclusion that made May's paper so important. May drew attention to the fact that very simple models can have very complex behaviour. The reverse is also important to note — just because something has apparently complex behaviour it does not mean that the underlying causes of the behaviour are also complex.

The study of the logistic equation is an example of a case where a very simple model had a profound effect on people's understanding of how systems governed by simple laws could behave. This was not because it was an accurate model. The logistic equation was relatively easy to study and yielded much fascinating mathematics which caught the imagination of the mathematics community. This mathematics is all highly worthy of study in its own right. But more importantly for its impact, its underlying behaviour was generic and led to a better understanding of apparently complex behaviours observed elsewhere. However, one should never forget the basic purpose of any model is to provide an insight into the behaviour of your original problem.

7. The Ising model

The previous problem used a continuum model of an intrinsically discrete system — a population. We will now look at another intrinsically discrete system — atoms — but where we retain their discrete nature even though the numbers involved are huge. The Ising model is an attempt to understand why ferromagnetic materials such as iron can remain magnetic at lower temperatures, but lose their magnetism at some critical temperature — the Curie point or Curie temperature. The basic phenomenon is that each atom of iron acts like an individual bar magnet, which can have arbitrary orientation. At low temperatures there is a tendency for the magnetic fields of nearby atoms of iron to align themselves in the same direction, thus creating an additive effect which generates the large-scale magnetic fields that we are familiar with. However, heating things excites the atoms in their structure and introduces a gradual tendency towards disorderly behaviour. Where does the abrupt behaviour come from?

In reality this is a complex problem in quantum mechanics, with the atoms of iron lying in a regular three-dimensional lattice — face-centred cubic, which is the densely packed arrangement of oranges seen on a green-grocer's barrow. We can try and understand this relatively complex system by looking at simpler versions and seeing if they share any of the properties we are interested in. Following Ising [12], for a first attempt at simplification we will make the problem one-dimensional, and look at a string of equally spaced atoms as shown in Fig. 7. Allowing general orientations of the magnetic moments of each atom is still a bit difficult, so we will restrict the magnetic moments of the atoms to pointing straight up or down as shown in Fig 8. We now want to investigate how the tendency of these atoms to align themselves due to inter-atomic interactions balances the effect of heating the system, which tends to jumble up the orientations of the atoms. Our experience in playing with magnets would lead us to think that the tendency would be for adjacent magnets to want to align themselves in opposite directions, with one atom's north pole as close as possible to its

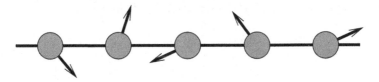

Fig. 7.   A one-dimensional string of iron atoms, with arbitrarily aligned magnetic fields.

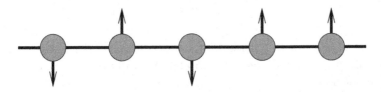

Fig. 8.   A one-dimensional string of iron atoms, with magnetic fields pointing up or down.

neighbour's south pole. However, on the atomic level this is not the case and iron atoms tend to want to align themselves pointing in the same direction.

We need some way to represent the state of our system. We can represent the string of atoms shown in Fig. 8 by a sequence of ± 1s. For example, in the case shown,

$$\ldots, -1, +1, -1, +1, +1, \ldots$$

we will let the ith element in this sequence be S_i.

In order to use the tools of statistical mechanics, we define the Hamiltonian for the system, $\mathcal{H}(S)$. This is essentially a measure of the energy. We will make some further assumptions. Firstly we note that, for example, the magnetic field associated with an atom is a dipole, whose field strength decays relatively rapidly. Hence, the interactions of an iron atom with its nearest neighbours will be significantly stronger that the direct interactions with atoms further away. In our model, we will assume that the interactions with each atom's nearest neighbours are the only important ones.

As adjacent iron atoms like to have their magnetic fields aligned, we will take the energy associated with S_i and S_{i+1} being aligned to be -1 (low energy — preferred) and in opposition $+1$ (high energy) in some appropriate units of energy. We can then write this interaction energy as

$$-S_i S_{i+1}.$$

The total energy due to all the interactions is then

$$\mathcal{H}(S) = -\sum_i S_i S_{i+1}. \tag{31}$$

The system of atoms can be in a vast number of different states. From statistical mechanics, the probability of being in any given state is proportional to

$$e^{-\beta \mathcal{H}(S)}, \tag{32}$$

where $\beta = 1/kT$, where k is the Boltzmann constant and T is the temperature. Given we have not been too careful about our units, we can assume that in our units $k = 1$.

What happens as we change β? There is a competition here between the low energy levels of ordered states, which individually have a high probability of being observed, and disordered states which individually have a lower probability but there are far more of them. As the temperature increases the importance of the energy difference decreases, and it is the large number of more random states that is important. As the temperature decreases it is the difference in the energy levels that predominates. This argument just tells us that a system at a higher temperature is more likely to look disordered than one at a lower temperature. However, it does not say anything about the transition. In summary:

- If β is small (T is large — hot) then all states will be roughly equally likely. The number of disordered states is vastly greater than the number of ordered states, so we would typically see more disordered states.
- If β is large (T is small — cold) then systems with low values of $\mathcal{H}(S)$ will be favoured. These states occur when many of the adjacent atoms are aligned, so we would typically see more ordered states.

Unfortunately, Ising was able to show that for the one-dimensional model there was no Curie point: the transition from disordered states to ordered states was smooth.

An obvious way of adding complexity to such a model which will make it closer to our original problem is to extend it to two dimensions, and to consider a system where the atoms are located on a square lattice. If we again restrict ourselves to the energy being only due to the interactions with nearest neighbours, then we get an energy of a given arrangement

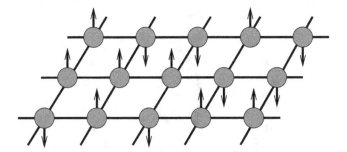

Fig. 9. A two-dimensional square array of iron atoms, with magnetic fields pointing up or down.

being given by the Hamiltonian for the system:

$$\mathcal{H}(S) = -\sum_{i,j} S_{i,j}S_{i+1,j} + S_{i,j}S_{i,j+1}.$$

It was shown by Onsager [13] that there is a critical point of this system.

We have seen that the probability of a given configuration, S, is proportional to $e^{-\beta\mathcal{H}(S)}$, so the probability of any given configuration is

$$\frac{e^{-\beta\mathcal{H}(S)}}{\sum_{S'} e^{-\beta\mathcal{H}(S')}} = \frac{e^{-\beta\mathcal{H}(S)}}{\mathcal{Z}}, \tag{33}$$

where the summation in the denominator is over all possible configurations, S'. The function in the denominator

$$\mathcal{Z} = \sum_{S'} e^{-\beta\mathcal{H}(S')} \tag{34}$$

is called the partition function.

If there is to be any change in the behaviour of a system it must be reflected in a change in the probabilities of the configurations. The numerator of the probability of each configuration is well behaved with no sudden transitions, so we must look at the partition function.

Onsager showed that for the two-dimensional Ising model with N atoms the partition function is approximated by

$$\mathcal{Z} \approx \prod_{t=1}^{N} \left(\left(z + z^{-1}\right)^2 - 4\left(z - z^{-1}\right)\cos(2\pi t/N) \right), \tag{35}$$

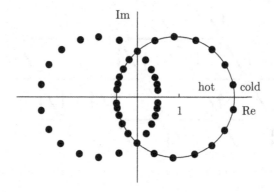

Fig. 10.   Schematic diagram of the zeros of the partition function in the complex z-plane.

where

$$z = e^{2\beta}. \tag{36}$$

If we look for the zeros of this function we can look for the zeros of the factors

$$f(z) = \left(z + z^{-1}\right)^2 - 4\left(z - z^{-1}\right)\cos\left(2\pi t/N\right)$$

$$= z^{-2}\left(\left(z^2 + 1\right)^2 - 4z\left(z^2 - 1\right)\cos\left(2\pi t/N\right)\right). \tag{37}$$

For large grids, the zeros of $\mathcal{Z}$ and f tend to lie on two circles with centres ± 1 passing through $\pm i$, and have radii $\sqrt{2}$ shown schematically in Fig. 10. Since the temperature, T, is positive, so is $\beta = 1/T$, and so we are only interested in values of e^β greater than 1. As z increases along the real axis it passes through the ring of zeros. At this point there is the possibility of a shift in behaviour. This occurs when $z = 1 + \sqrt{2}$, so

$$\beta = \frac{\log\left(1 + \sqrt{2}\right)}{2} \approx 0.4407 \qquad \text{or} \qquad T_c = \frac{2}{\log\left(1 + \sqrt{2}\right)}. \tag{38}$$

Some numerical simulations of the Ising model made using the Metropolis algorithm are shown in Fig. 11. Cells where the magnetic field points up are white, and those where they point down are black. The first image, for $\beta = 0.3407$, is for a temperature just above the Curie point. Although there is some level of very short-range correlation, over large-scales the magnetic field averages to zero. The last image is for $\beta = 0.5407$, just below the Curie temperature. Here, almost all the atoms have a magnetic field which

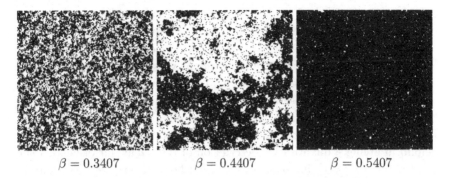

$$\beta = 0.3407 \qquad \beta = 0.4407 \qquad \beta = 0.5407$$

Fig. 11. Simulations of the Ising model done using the Metropolis Algorithm on 256×256 grid for supercritical ($\beta = 0.3407$), critical ($\beta = 0.4407$) and subcritical ($\beta = 0.5407$) temperatures.

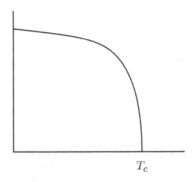

T_c

Fig. 12. Mean magnetic field as a function of temperature.

points down. It could equally likely have ended up with them all pointing up. There are still many individual atoms pointing in the wrong direction, but their proportion is small. At the Curie point, $\beta = 0.4407$, although the field is approximately zero on average, the structure of the regions of different magnetic orientations becomes more complex, with evolving patches which predominantly have magnetic fields pointing up or down appearing on all scales. Further analysis shows that the mean magnetic field predicted by the Ising model just below the Curie point is proportional to

$$(T_c - T)^{1/8}. \tag{39}$$

Is this an accurate model for ferromagnetism in iron? The obvious answer is "no". There are several clear problems with the model. Firstly, this is a two-dimensional model while a lump of iron is intrinsically three-dimensional. In this model, there are only four nearest neighbours, while in solid iron each atom has 12 nearest neighbour, twice the number that would be present in the obvious three-dimensional version of the Ising model. There will also be an effect of ignoring other atoms that are not nearest neighbours, although these are further away in face-centred cubic latices than in the simple three-dimensional Ising model. Does this mean that the Ising model has no merit? This is definitely not the case. It may not be helpful in giving a quantitative description of the behaviour of magnetism in iron, but it is very important in showing how the critical behaviour of magnetism at the Curie temperature can come out of a relatively simple model. This critical behaviour does not have to be built into the model, but is something that emerges from the underlying interactions of the atoms. Just as for the logistic equation, it is this insight that makes this model so important. Quite complex behaviour can emerge from simple models.

8.   Earthquakes and the tail of distributions

Some random phenomena in the natural world have odd distributions. If you look at the frequency of earthquakes you find that earthquakes of magnitude 6 on the Richter scale occur roughly 10 times more frequently than earthquakes of magnitude 7, which in turn are 10 times more frequent than earthquakes of magnitude 8, and so on. Since the Richter scale is logarithmic, this means that the frequency of earthquakes decreases algebraically with the energy. The tail of the distribution decays far more slowly than, say, the tail of the normal distribution.

There are other physical systems that have such an algebraic decay in the frequency when extreme events are considered. In a paper by Bak *et al.* [14] a model was put forward with such behaviour. This model starts off by looking at sand falling down a slope. It first considers a simple one-dimensional model with stacks of cubic grains located in a line. If the height difference between two adjacent stacks exceeds some threshold, then a grain falls from the higher stack to the lower stack. This means that the height difference is reduced by 2. At the same time the height differences between these two stacks and their other neighbours are both increased by 1. A schematic diagram of the process is shown in Fig. 13. The height of the grains of sand are set to 1 in some appropriate units. If a step is above some critical value, z_c, then a grain falls down the step. If the grain at n

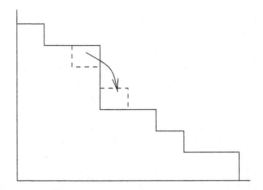

Fig. 13. When a step becomes too high, a "grain" falls down the step.

falls down, then the step initially to the right of the falling grain decreases in height by 2, and the adjacent steps to left and right both increase by 1. We can model this as

$$z_n \rightarrow z_n - 2,$$
$$z_{n\pm 1} \rightarrow z_{n\pm 1} + 1, \tag{40}$$

where z_n is the height of a step. This process may make one or both of the adjacent steps unstable. If so, then they will collapse in the same way. This process is repeated until all the steps are below the critical level. Thus, the collapse of a step at one point many lead to a cascade with many grains falling down.

Bak *et al.* [14] had two models for inducing possible cascades:

(1) Adding a grain of sand at a random pile, so the step to the left decreases by 1, and the step to the right increases by 1:

$$z_n \rightarrow z_n + 1,$$
$$z_{n-1} \rightarrow z_{n-1} - 1. \tag{41}$$

(2) Imagine gradually tilting the slope to make it steeper, then at some point a grain will fall. This is simulated by adding height 1 to one step at random. That is at some random step n

$$z_n \rightarrow z_n + 1. \tag{42}$$

As with the one-dimensional Ising model, this one-dimensional model is not very exciting. You get a state where each step is just below the critical

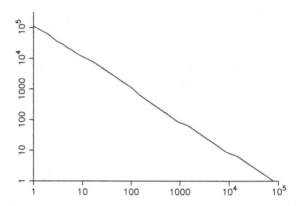

Fig. 14. A log–log plot of the frequency of the simulated cascades as a function of their size, using the model from (43).

value. You add a grain of sand and a cascade starts with a grain of sand falling to the bottom edge of the slope, still leaving the slope just below critical.

The authors extended the second model in a mathematically obvious way to two- and three-dimensions in space. For example, in two dimensions if at any point on a square grid $S_{i,j} \geq 4$ then

$$S_{i,j} = S_{i,j} - 4,$$

$$S_{i\pm 1,j} = S_{i\pm 1,j} + 1, \quad S_{i,j\pm 1} = S_{i,j\pm 1} + 1. \tag{43}$$

As before, if the adding of 1 to adjacent points makes them unstable then these in turn collapse, and cascades may form.

How the above would be derived from a model of a slope of sand, and what each $S_{i,j}$ corresponds to, is not quite so clear. However, when investigated numerically it is found that the sizes of the landslides caused by the the addition of grains of sand did have an algebraic tail to their frequency distribution as shown in Fig. 14.

This model may not be very realistic for a pile of sand, but perhaps would be better as a simple model for earthquakes at a fault line. If you imagine the surfaces of two rough bodies of rock rubbing against each other at a fault line then patches of the joints between the two rock masses will have different loads on them. As the rock masses move slowly past each other, the stresses at the interface will gradually increase, and not necessarily uniformly. Eventually at some point the stresses will exceed some maximum yield stress and the rocks will slip past each other, releasing

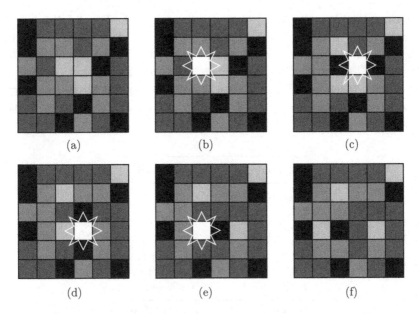

Fig. 15.   Stress levels of squares are white — 4 and black — 0, with lighter shades of grey having higher stresses. The diagrams shows (a) initial configuration, (b) addition stress added to top left central square which cracks, and (c)–(f) show subsequent stress redistribution and cracking, with four patches cracked overall.

much of the stresses at that location. Because the total stress is essentially conserved over larger scales, the released stresses will be taken up by other nearby patches of rock. This process could be modelled by the above scheme. The above process is illustrated in Fig. 15. We start in (a) with an array of patches with random stresses, all less than the threshold level of 4. As stresses build up non-uniformly occasionally we add to the stress of a random point. Occasionally one patch will be overloaded as shown by the white patch in (b). This patch then cracks, releasing its stress which is distributed evenly over its four nearest neighbours as shown in (c). This takes one of the adjacent patches up to breaking point, which in turn cracks and releases its stresses, and so on (d)–(f). This process is repeated until all patches are below the fracture point. In this case the initial added stress resulted in four patches cracking.

The above process can be modelled numerically on a grid. At the edge of the region the stresses are absorbed, so the system will come to rest eventually after each added stress unit. The size of an "earthquake" is determined to be the number of patches that fracture after each stress addition, ignoring possible multiple cracking of the same patch. The distribution of "earthquakes" found by this model has a power-law distribution shown in Fig. 14, with the probability of a quake of the size N being proportional to N^α with $\alpha \approx 1$.

A table taken from U.S. Geological survey web pages of the frequency of occurrence of earthquakes is shown below, showing some agreement with this model.

Magnitude	Average Annually
8 and higher	1[*]
7–7.9	17[†]
6–6.9	134[†]
5–5.9	1319[†]
4–4.9	13,000 (est.)
3–3.9	130,000 (est.)
2–2.9	1,300,000 (est.)

[*]Based on observations since 1900
[†]Based on observations since 1990

This model is a very simple model of earthquakes. There are better ones that reflect a more realistic version of the dynamics, for example, the model of Carlson and Langer [15]. However, even this simple model does get an approximation to the large amplitude decay to the distribution of earthquakes.

It should be noted, however, that this model does not have any obvious physical connection to the applications such as cosmic strings, turbulence or coastal lines mentioned by Bak *et al.* [14]. In the context of these problems that motivated their study, this model would be of the weakest kind, where there is a similarity in the behaviour of the model and the phenomena of interest, but no obvious connection between the underlying causes and the model itself.

9. Dynamical systems and chaotic behaviour

Dynamical systems theory has had a significant impact on applied mathematics and mathematical modelling. We will look at two dynamical systems that have had a notable impact in different ways. The first is the Lorenz system discovered by Edward Lorenz [16], who coined the phrase "the butterfly effect" to describe the sensitivity of its chaotic behaviour to variations in initial conditions. The second is the Rikitake model [17], which gave an insight into the occasional reversals of the Earth's magnetic field.

The Lorenz equations [16] is a set of three coupled ordinary differential equations

$$\frac{dx}{dt} = \sigma(y - x), \tag{44}$$

$$\frac{dy}{dt} = \rho x - xz - y, \tag{45}$$

$$\frac{dz}{dt} = xy - \beta z. \tag{46}$$

Frequently the choice of parameters is that $\sigma = 10$ or 7 and $\beta = 8/3$. These are fixed. The behaviour as ρ is varied is then investigated. This choice of fixed and varied parameters and the value of the fixed parameters is rarely explained.

The Lorenz equations originally came from studying a model of convection in the atmosphere. Lorenz observed irregular behaviour in his original model — chaos. When he tried to reproduce the results by re-entering numbers from a previous run he got behaviour that diverged — sensitivity to initial conditions. These two properties are ubiquitous in the many chaotic systems that were subsequently discovered.

The Lorenz equations are a simplified version of his original model. They can be derived from the governing equations for convection between two horizontal boundaries with an imposed vertical temperature difference, as shown in Fig. 16. The relevant governing equations are the Navier–Stokes equation, the equation of continuity and the heat equation:

$$\frac{\partial \mathbf{u}}{\partial t} + \mathbf{u} \cdot \nabla \mathbf{u} = -\frac{1}{\rho_0}\nabla p + g\alpha T\hat{\mathbf{y}} + \nu\nabla^2\mathbf{u}, \tag{47}$$

$$\nabla \cdot \mathbf{u} = 0, \tag{48}$$

$$\frac{\partial T}{\partial t} + \mathbf{u} \cdot \nabla T = \kappa\nabla^2 T. \tag{49}$$

Here, we have made the Boussinesq approximation — the density variations are neglected except in the buoyancy term of the Navier–Stokes equation. This requires that the temperature difference across the boundaries is not

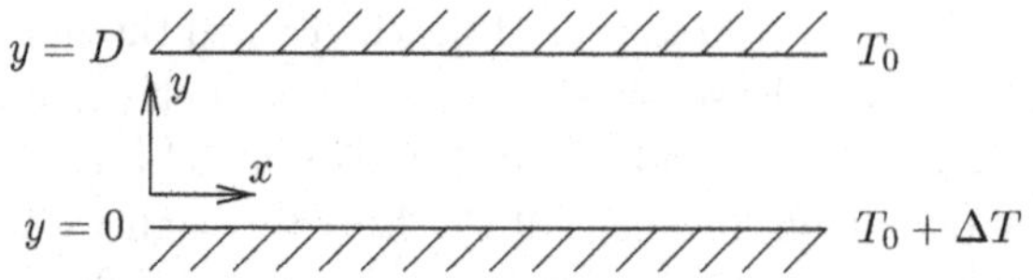

$$y = D \quad \text{/////////////} \quad T_0$$

$$y = 0 \quad \text{/////////////} \quad T_0 + \Delta T$$

Fig. 16. Schematic diagram of convection between two horizontal boundaries.

too large. The density is given by

$$\rho = \rho_0(1 - \alpha(T - T_0)) \tag{50}$$

and the boundary conditions are

$$T = T_0, \quad v = 0 \quad \text{on } y = D, \tag{51}$$

$$T = T_0 + \Delta T, \quad v = 0 \quad \text{on } y = 0. \tag{52}$$

In reality, we should probably apply the no-slip boundary condition $u = 0$ on $y = 0$ and $y = D$, but often stress-free boundary conditions are used, which makes the analysis easier.

A check on the dimensional quantities involved yields

$$\nu \sim L^2 T^{-1}, \qquad \kappa \sim L^2 T^{-1}, \qquad g \sim LT^{-2},$$

$$\rho_0 \sim ML^{-3}, \qquad \Delta T \sim \theta, \qquad \alpha \sim \theta^{-1}, \qquad D \sim L.$$

There are seven parameters involving four basic quantities, so we should have three non-dimensional parameters. But g only appears when multiplied by α, and so we have another restriction that reduces the problem to two non-dimensional quantities. Conventionally, these are the Rayleigh number and the Prandtl number

$$\text{Ra} = \frac{g\alpha\Delta T D^3}{\nu\kappa}, \qquad \sigma = \frac{\nu}{\kappa}. \tag{53}$$

Other choices could be made, but again it is not usually a good idea to fight against convention in cases like this. Making alternative choices will be likely to alienate others without any clear benefit.

Non-dimensionalising the equations using scalings of length $\sim D$, time $\sim D^2/\kappa$, temperature $\sim \Delta T$, speed $\sim \kappa/D$, pressure $\sim \rho_0\kappa\nu/D^2$, gives the non-dimensional equations

$$\frac{1}{\sigma}\left[\frac{\partial \mathbf{u}}{\partial t} + \mathbf{u}\cdot\nabla\mathbf{u}\right] = -\nabla p + \text{Ra}T\hat{\mathbf{y}} + \nabla^2\mathbf{u}, \tag{54}$$

$$\nabla \cdot \mathbf{u} = 0, \tag{55}$$

$$\frac{\partial T}{\partial t} + \mathbf{u} \cdot \nabla T = \nabla^2 T, \tag{56}$$

where all the variables are now dimensionless. These have the simple conduction solution

$$\mathbf{u} = \mathbf{0}, \qquad T = 1 - y. \tag{57}$$

In experiments, when the temperature difference is small this is the solution that is observed. When the temperature difference is sufficiently large convection starts — the fluid starts moving. It can be shown that between parallel plates the onset of convection is usually in the form of parallel convection rolls. If we look at a two-dimensional version of the equations in a plane perpendicular to the roll axes then we can use a streamfunction, ψ, such that

$$u = \frac{\partial \psi}{\partial y}, \qquad v = -\frac{\partial \psi}{\partial y}, \tag{58}$$

to represent the flow. If we take T' to be the deviation from the conduction solution then the governing equations, after taking the curl of the momentum equation, become

$$\frac{1}{\sigma} \left[\frac{\partial \nabla^2 \psi}{\partial t} + \psi_y \nabla^2 \psi_x - \psi_x \nabla^2 \psi_y \right] = \mathrm{Ra} T'_x - \nabla^4 \psi, \tag{59}$$

$$\frac{\partial T'}{\partial t} + \psi_y T'_x - \psi_x T'_y + \psi_x = \nabla^2 T'. \tag{60}$$

If we neglect the nonlinear terms we can look for solutions of the form

$$\psi = \psi(y) \sin \alpha x \, e^{\lambda t}, \qquad T' = T(y) \cos \alpha x \, e^{\lambda t}. \tag{61}$$

We end up with an eigenvalue problem involving a fourth-order ordinary differential equation for $\psi(y)$ and a second-order ordinary differential equation for $T(y)$. If we use stress-free boundary conditions we find both $\psi(y)$ and $T(y)$ have a possible y-dependency of the form $\sin n\pi y$, for $n = 1, 2, \ldots$. The eigenvalue problem then reduces to the algebraic problem

$$\left[\frac{\lambda}{\sigma} + (\alpha^2 + n^2\pi^2) \right] \left[\lambda + (\alpha^2 + n^2\pi^2) \right] (\alpha^2 + n^2\pi^2) = \alpha^2 \mathrm{Ra}. \tag{62}$$

For steady onset of convection (as opposed to oscillatory onset) $\lambda = 0$ and

$$\text{Ra} = \frac{(\alpha^2 + n^2\pi^2)^3}{\alpha^2}. \tag{63}$$

This has a minimum when $n = 1$ and $\alpha = \pi/\sqrt{2}$, and so gives the critical value of the Rayleigh number

$$\text{Ra}_c = \frac{27\pi^4}{4}. \tag{64}$$

Above this value there will always be unstable convection modes. These grow without bound, and will eventually violate the linearity approximation.

The nonlinear behaviour of the fluid when the heating is just above this marginal threshold can be investigated using a weakly-nonlinear analysis. We will not go into the details here. In essence we look for solutions near marginal stability where we expect solutions to be small. We pose expansions of the form

$$\psi = \epsilon\psi_0 + \epsilon^2\psi_1 + \epsilon^3\psi_2 + \cdots, \tag{65}$$

$$T = \epsilon T_0 + \epsilon^2 T_1 + \epsilon^3 T_2 + \cdots, \tag{66}$$

$$\text{Ra} = \frac{27\pi^4}{4} + \epsilon\text{Ra}_1 + \epsilon^2\text{Ra}_2 + \cdots. \tag{67}$$

In this analysis we find $\text{Ra}_1 = 0$ and $\text{Ra}_2 > 0$. We also find $\psi_1 = 0$ and $T_1 \propto \sin 2\pi y$. This gives rise to the relation between the Rayleigh number and the amplitude shown in Fig. 17. This reflects the fact that convection tends to make the interior of the fluid more evenly mixed, thus reducing the effective core temperature gradient and the effective Rayleigh number.

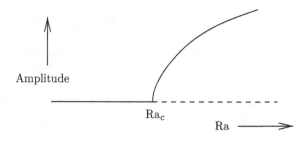

Fig. 17. Dependency of the amplitude of the convection rolls as a function of the Rayleigh number, Ra.

This analysis can be extended by looking for the time variation of these convection rolls to find out how the system evolves. A simple way of looking at this is to assume

$$\psi = a(t) \sin \alpha x \sin \pi y, \tag{68}$$

$$T = b(t) \cos \alpha x \sin \pi y + c(t) \sin 2\pi y. \tag{69}$$

These are substituted into the governing equations (59) and (60). Where the nonlinear interactions generate other terms of a different spatial form they are neglected. This yields a set of equations

$$\frac{da}{dt} = \sigma \left[\frac{\sqrt{2}Ra}{3\pi} b - \frac{3\pi^2}{2} a \right], \tag{70}$$

$$\frac{db}{dt} = \frac{\pi}{\sqrt{2}} a - \frac{3\pi^2}{2} b - \frac{\pi^2}{\sqrt{2}} ac, \tag{71}$$

$$\frac{dc}{dt} = \frac{\pi^2}{2\sqrt{2}} ab - 4\pi^2 c. \tag{72}$$

where we use $\alpha = \pi/\sqrt{2}$. With some rescaling of t and the other variables this reduces to the Lorenz equations

$$\frac{da}{dt} = \sigma \left(b - a \right), \tag{73}$$

$$\frac{db}{dt} = \rho a - b - ac, \tag{74}$$

$$\frac{dc}{dt} = ab - \frac{8}{3} c, \tag{75}$$

where $\rho = Ra/(27\pi^4/4)$. Note that the the meaning of two of the original parameters are that we saw in the original Lorenz equations (44) and (46) are now clearer. The first parameter, σ, is just the Prandtl number. This is typically taken to be either 7 or 10 for water. The second parameter, β, is derived from the aspect ratio of the marginally unstable convection rolls and takes the value 8/3. The last parameter, ρ, is the ratio of the applied temperature difference to that at marginal stability. In experiments, the Prandtl number and the geometry tend to be fixed, and the temperature difference is varied. The equivalent here is to fix σ and β and to vary ρ, hence the choice of fixed and variable parameters in other studies.

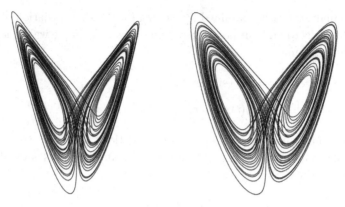

Fig. 18.   Stereogram of the Lorenz attractor for $\rho = 28$, the path taken by a solution of the Lorenz equations.

Fundamental to the derivation of these equations is the assumption of small amplitudes and the associated restriction that ρ is just over 1. In this case, the solution of these equations gives an accurate representation of the behaviour of a heated layer of fluid. If you use large values such as $\rho = 28$ you get interesting behaviour such as that shown in Fig. 18. The results are all very pretty, but not a lot to do with thermal convection. At these levels of heating you would expect convection to be three-dimensional and turbulent, and nothing like the rolls of the underlying model.

Other dynamical systems have had a significant impact in other areas. For example, Rikitake [17] proposed a model of coupled dynamos that mimic a mechanism for the generation of the Earth's magnetic field. As we saw earlier, permanent magnets cannot exist above the Curie temperature, and the core of the Earth is too hot for this to be the origin of the Earth's magnetic field. Instead the magnetic field is generated by the convective motions in the Earth's core, see for example Moffatt [18]. One observation that has been made from geological samples is that around 400 times in the last 330,000,000 years the polarity of the Earth's magnetic field has reversed, with the magnetic north and south poles swapping position.

The convection in the molten metallic core can generate magnetic fields spontaneously, but it needs other inputs such as rotation. This can give rise to a magnetic field which has a strong toroidal element to the mean field in the core (which essentially goes around the Earth's axis, and which we do not see on the surface), and a poloidal part (the part that does not go around the axis) that comes out of surface and with which we are familiar.

The convection in the core interacts with the toroidal field to generate the poloidal field, and then generates the toroidal field from its interaction with the poloidal field.

The Rikitake model looks at an idealised system where there are two interacting dynamos. The current generated by each is used to provide the magnetic field in the other dynamo. The model ends up with four coupled non-dimensional equations:

$$\frac{dX}{dt} + \mu X = ZY, \tag{76}$$

$$\frac{dY}{dt} + \mu Y = ZX, \tag{77}$$

$$\frac{dZ}{dt} = \frac{dV}{dt} = 1 - XY, \tag{78}$$

where X and Y represent the strengths of the two magnetic fields (which can be taken to represent the poloidal and toroidal fields), and Z and V the rotation rates of the dynamos.

When the driving forces in this model are weak then no magnetic field is generated. However, when there is sufficient input of energy magnetic fields can be spontaneously generated, and the fields generated can be chaotic. Such chaotic behaviour is shown in Fig. 19. This trajectory of this solution is seen to spiral around two different points. This shows that the magnetic fields oscillates around two different values, with occasional transition between these values. This represents an occasional flipping of the overall polarity of the magnetic field. This is not a very realistic model, but it does show that spontaneous field reversal in a system that generates its own magnetic field is possible. This behaviour has been subsequently observed in other models of the generation of magnetic fields by the dynamo effect in the Earth and in the Sun [18].

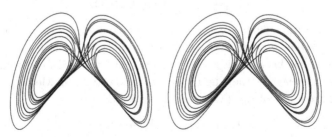

Fig. 19.   Stereogram of the solutions of the Rikitake equations for $\mu = 1$, $K = 2$.

10.   Other models and conclusions

The models presented here are very much a personal choice based on models that I have encountered and made an impact on me. There are other examples of models that could have been chosen. For example, we could have looked at the occurrence of Fibonacci numbers in the spirals of sunflowers and other plants, which is often found puzzling. Vogel [19] showed that if you place sunflower seeds near the centre with a regular angular spacing of $2\pi * \phi$, where ϕ is the golden ratio $\phi = (\sqrt{5} - 1)/2$, and allow them to push outwards, then the seeds are very efficiently packed, and the number of spirals that you observe in the seed head will all be Fibonacci numbers. But no mechanism is offered for how sunflowers select such an angle, and other angles yield packing that is almost as efficient, but do not yield the right numbers of spirals. However, Douady and Couder [20] offered a simple model where new seeds were placed near the centre in positions that were, in some sense, as far away from the other existing seeds as possible. Under a wide range of parameter values, this model naturally settled down to a configuration where the spirals observed matched the Fibonacci pattern. This mechanism seems to be reasonably robust, and the argument that this may be an underlying mechanism for the spiral formation seems far more compelling. Although the Vogel approach produced the right answer, it did not seem to present a robust mechanism that would deliver the spirals so frequently observed.

The study of combustion is very important in many ways, for example one may be interested in enhancing the efficiency of engines or in reducing the levels of pollution. When looking at problems in combustion you can get buried in the details of the chemical reactions. For example, we learn at school the chemical reaction for the burning of methane with oxygen is given by

$$CH_4 + 2O_2 \rightarrow CO_2 + 2H_2O.$$

A simple reaction where oxygen in the air combines with the methane to create carbon dioxide and water. But reality is a bit more complex. An attempt has been made to catalogue all the reactions involved in the burning of methane in the The Leeds Methane Oxidation Mechanism. The following are a few of the 174 reactions involved:

$$CH_4 + O_2 \leftrightarrow CH_3 + HO_2,$$
$$H_2 + CH_2 \leftrightarrow CH_3 + H,$$

$$H_2 + O \leftrightarrow OH + H,$$
$$H_2O + H \leftrightarrow H_2 + OH,$$
$$CH_4 + C \leftrightarrow CH + CH_3.$$

The complete set of reactions involve a total of 35 different molecules. This ignores possible reactions with nitrogen, which could be a source of pollution. Of course this is a simple case! Reactions for the combustion of petrol in cars (mostly octane — C_8H_{18}) are much more complex.

Trying to incorporate the full chemistry in a model of, for example, combustion in an engine is difficult as the reaction rates of all these individual reactions can differ by several orders of magnitude, and to resolve all of them would be tricky and time consuming. An approach pioneered by Stephen Pope and Ulrich Maas [21, 22] uses the ideas behind manifold theory in dynamical systems to collapse the complex chemistry onto a smaller dimension subspace where the chemistry evolves in a more manageable way. When I first came across this approach towards the end of the last century, I was aware that there were chemists and engineers in the audience who did not find the elegance of the mathematics compelling, and who were convinced that no good would come from this approach. Fortunately, there is now a higher level of acceptance of this method. The moral of this is that you should be aware of your audience: not everyone is going to be impressed by the elegance of your sophisticated mathematical techniques, and to get your point across to your intended audience you may need to adapt your approach or presentation. You may also have to persevere.

Whilst personally working on a problem in combustion there was a desire for a realistic steady flow that would mix converging flows of fuel and air. Due to a misunderstanding of an idea put forward by my then supervisor, I followed up my ideas and we ended up discovering a new family of steady state solutions of the Navier–Stokes equation [23]. The existence of these theoretical solutions was later verified in real fluid flows by Andreotti *et al.* [24]. Simplified models can throw up ideas that subsequently turn out to yield real results, even if not for the original problem. It can be worthwhile following up unexpected results and ideas, although deliberately misunderstanding your supervisor is probably not a good idea in general.

We have focused here on models that have involved a significant degree of simplification, and have yielded results that have often had a significant impact. In all cases it should be remembered that this simplification may come at the cost of quantitative accuracy, but with the advantage that the underlying processes are clearer. When a higher degree of accuracy is required then the scope for simplification may be reduced, although the use of non-dimensional parameters to reduce the size of the problem will nearly always be worthwhile. How much simplification is required or acceptable will always be a matter of judgment, and may depend on those who will use your model in the future.

A mathematical model of a problem is rarely an isolated and totally original piece of work. It is more often a development based on the foundations laid by others, or the application of the ideas from other areas to a new problem. It is always a good idea to be familiar with different models and approaches from a wide range of areas.

Modelling — what is it good for? An awful lot — you can provide realistic simulations of real phenomena, you can develop rational simplifications to complex problems that give insight into the underlying phenomena and you may end up doing really interesting mathematics. But if you want to make headlines in the national media...

References

[1] A. C. Fowler, *Mathematical Models in the Applied Sciences*, Cambridge University Press, Cambridge, 1997.
[2] E. Cumberbatch and A. Fitt, *Mathematical Modelling — Case Studies from Industry*, C.U.P., Cambridge, 2001.
[3] J. Ockendon, S. Howison, A. Lacey and A. Movchan, *Applied Partial Differential Equations*, O.U.P., Oxford, 1999.
[4] L. Bégue, B. J. Bushman, O. Zerhouni, B. Subra and M. Ourabah, 'Beauty is in the eye of the beer holder': People who think they are drunk also think they are attractive, *J. Psych.*, **104** (2012), 225–234.
[5] J. T. Jeong and H. K. Moffatt, Free-surface cusps associated with flow at low reynolds number, *J. Fluid Mech.*, **276** (1994), 307–325.
[6] E. Buckingham, On physically similar systems; illustrations of the use of dimensional equations, *Phys. Rev.*, **4** (1914), 345–376.
[7] G. I. Taylor, The formation of a blast wave by a very intense explosion. Technical Report RC-210, Civil Defence Research Committee, 1941.
[8] G. I. Taylor, The formation of a blast wave by a very intense explosion. I, Theoretical discussion, *Proc. Roy. Soc.*, **A201** (1950), 159–174.
[9] G. I. Taylor, The formation of a blast wave by a very intense explosion. II, The atomic explosion of 1945, *Proc. Roy. Soc.*, **A201** (1950), 175–186.
[10] M. J. Lighthill and G. B. Whitham, On kinematic waves. II. A theory of traffic flow on long crowded roads, *Proc. Roy. Soc.*, **229** (1955), 317–345.
[11] R. M. May, Simple mathematical models with very complicated dynamics, *Nature.*, **261** (1976), 459–467.

[12] E. Ising, Beitrag zur theorie des ferromagnetismus, *Z. Phys.*, **31** (1925), 253–258.

[13] L. Onsager, Crystal statistics I. A two-dimensional model with an order-disorder transition, *Phys. Rev.*, **65** (1944), 117–149.

[14] P. Bak, C. Tang and K. Wiesenfeld, Self-organized criticality: An explanation of $1/f$ noise, *Phys. Rev. Lett.*, **59** (1987), 381–384.

[15] J. M. Carlson, J. S. and Langer, Mechanical model of an earthquake fault, *Phys. Rev. A.*, **40** (1989), 6470–6484.

[16] E. N. Lorenz, Deterministic nonperiodic flow, *J. Atmos. Sci.*, **20** (1963), 130–141.

[17] T. Rikitake, Oscillations of a system of disk dynamos, *Proc. Camb. Phil. Soc.*, **54** (1958), 89–105.

[18] H. K. Moffatt, *Magnetic Field Generation in Electrically Conducting Fluids.*, Cambridge University Press, Cambridge, 1983.

[19] H. Vogel, A better way to construct the sunflower head, *Math. Biosci.*, **44** (1979), 179–189.

[20] S. Douady and Y. Couder, Phyllotaxis as a physical self-organized growth process, *Phys. Rev. Lett.*, **68** (1992), 2098–2101.

[21] S. B. Pope, Computations of turbulent combustion: Progress and challenges, *Proc. Comb. Inst.*, **23** (1990), 591–612.

[22] V. Bykov and U. Maas, The extension of the ildm concept to reaction-diffusion manifolds, *Combust. Theor. Model.*, **11** (2007), 839–862.

[23] O. S. Kerr and J. W. Dold, Periodic steady vortices in a stagnation point flow, *J. Fluid Mech.*, **276** (1994), 307–325.

[24] B. Andreotti, S. Douady and Y. Couder, An experiment on two aspects of the interaction between strain and vorticity, *J. Fluid Mech.*, **444** (2001), 151–174.

Chapter 4

Finite Elements

Matthias Maischak

Department of Mathematics, Brunel University, Kingston Lane,
Uxbridge UB8 3PH, UK
matthias.maischak@brunel.ac.uk

This chapter is concerned with the numerical treatment of partial differential equations using the Finite Element Method (FEM). We will start with the concept of variational formulation, introduce the Galerkin method, discrete (polynomial) spaces, uniform, algebraically graded and adaptively refined meshes, prove some interpolation results, discuss *a priori* and *a posteriori* error estimates, and present several algorithms for implementing the FEM.

1. Introduction

In physics, engineering and other sciences many phenomena and processes can be described by partial differential equations (PDEs). Only in very special cases these PDEs can be solved exactly. For most applications, we need to solve these PDEs numerically, i.e., we have to compute approximate solutions and we need a way to determine the error of our approximation compared to the true (unknown) solution of the PDE. A well-known standard method to approximately solve many types of PDEs is the finite element method (FEM). In the following we will introduce the FEM in an abstract way, will give examples for PDEs and discrete spaces and discuss efficient ways to implement the FEM.

In Section 2, we introduce the conforming FEM and discuss *a priori* error estimates in a general sense. In Section 3, we prove interpolation results in one-dimensional, which will lead to *a priori* error estimates for h- and p-versions on different types of meshes. The results can be generalised to two-dimensional and three-dimensional problems. Section 4 is devoted to various algorithms needed in the implementation of the FEM. One major topic of interest is also the complexity of the algorithms. A low order of complexity is one of the main goals in developing an efficient FEM. Section 5 finally discusses an *a posteriori* error estimator which is easy to implement and a strategy for mesh-refinement resulting in an adaptive finite element method (AFEM). There are many textbooks on FEM available, see e.g., Braess [1], Brenner and Scott [2] or Johnson [3].

This chapter is based on a Finite Element Course for PhD students held by the author at the London Taught Course Centre in the years 2013 and 2014.

2. Finite element method

2.1. *Variational formulation*

There are many different types of PDEs, e.g., linear, nonlinear, with constant coefficients, with variable coefficients etc. As a model problem we will investigate the Poisson equation with homogenous boundary conditions

$$-\Delta u = f \text{ in } \Omega, \qquad u|_{\partial\Omega} = 0, \tag{1}$$

where $\Omega \subset \mathbb{R}^d$ is a simply connected domain, $\Delta = \sum_{i=1}^{d} \partial_{x_i}^2$ is the Laplace-operator and $f \in L^2(\Omega)$ is a given function. Δ is a second-order differential operator, therefore in the classical sense a solution of (1) has to be $u \in C^2(\Omega)$, i.e., the solution u has to be two times continuously differentiable. But in practice many problems do not have a classical solution. It is possible to generalise problem (1), by deriving the so-called variational formulation of (1), so that the new problem has always a unique solution. If the solution of the variational formulation is smooth enough it is also a solution of (1).

To derive the variational formulation of (1) we multiply (1) with a test function v and integrate over the domain Ω, i.e., we obtain

$$-\int_{\Omega} (\Delta u)v \, dx = \int_{\Omega} f(x)v(x) \, dx.$$

By observing $\Delta u = \text{div} \nabla u$ and applying the divergence theorem $\int_{\Omega} \text{div} \, \vec{w} \, dx = \int_{\partial\Omega} n\vec{w} \, ds$ we obtain

$$\int_{\Omega} f(x)v(x)dx = -\int_{\Omega} (\text{div} \nabla u)v \, dx = \int_{\Omega} \nabla u \nabla v \, dx - \int_{\Omega} \text{div}(v\nabla u) \, dx$$

$$= \int_{\Omega} \nabla u \nabla v \, dx - \int_{\partial\Omega} v \cdot n\nabla u \, ds. \tag{2}$$

We observe that (2) uses only first-order derivatives. This leads to the definition of the following function spaces.

Definition 1 (Sobolev spaces). Let $\Omega \subset \mathbb{R}^d$ be a simply connected domain. The Sobolev norms and spaces in d-dimensions are given by

$$\|u\|_{L^2(\Omega)}^2 = \int_\Omega (u(x))^2\, dx, \qquad L^2(\Omega) := \{u \,|\, \|u\|_{L^2(\Omega)} < \infty\},$$

$$\|u\|_{H^1(\Omega)}^2 = \|u\|_{L^2(\Omega)}^2 + \|\nabla u\|_{L^2(\Omega)}^2, \qquad H^1(\Omega) := \{u \,|\, \|u\|_{H^1(\Omega)} < \infty\},$$

$$H_0^1(\Omega) := \{u \in H^1(\Omega) \,|\, u|_{\partial\Omega} = 0\}.$$

Generally, we have

$$\|u\|_{H^n(\Omega)}^2 = \sum_{k=0}^n \sum_{\alpha_1+\ldots+\alpha_d=k} \|\frac{\partial^k u}{\partial_{x_1}^{\alpha_1}\ldots\partial_{x_d}^{\alpha_d}}\|_{L^2(\Omega)}^2.$$

The use of first-order derivatives in (2) indicates that $H^1(\Omega)$ is the proper function space. Taking into account the original boundary condition in (1) we see that the Sobolev space $H_0^1(\Omega) \subset H^1(\Omega)$ is the correct space. The variational formulation for the Poisson equation now reads:

Find $u \in H_0^1(\Omega)$ such that

$$\int_\Omega \nabla u(x)\nabla v(x)\, dx = \int_\Omega f(x)v(x)\, dx \quad \forall v \in H_0^1(\Omega). \tag{3}$$

2.2. *Hilbert space theory*

Let V be a Hilbert space, i.e., a vector space with inner product $(\cdot,\cdot)_V$, which is complete, i.e., every Cauchy sequence converges in V.

Let $a(\cdot,\cdot): V \times V \to \mathbb{R}$ be a bilinear form on V, with

$$a(v,v) \geq \alpha\|v\|_V^2 \quad \forall v \in V \text{ (ellipticity)}$$

and

$$a(u,v) \leq M\|u\|_V\|v\|_V \quad \forall u,v \in V \text{ (continuity)}.$$

Let $l(\cdot): V \to \mathbb{R}$ be a linear form on V, with

$$l(v) \leq M_l\|v\|_V \quad \forall v \in V \text{ (continuity)}.$$

The variational formulation reads:

Find $u \in V$ such that

$$a(u,v) = l(v) \quad \forall v \in V. \tag{4}$$

Theorem 1 (Lax–Milgram). [2, Theorem 2.7.7] *Let $a(\cdot, \cdot)$ be an elliptic and continuous bilinear form on V and $l(\cdot)$ be a continuous linear form. Then the variational formulation (4) has a unique solution $u \in V$ and there holds*

$$\|u\|_V \leq \frac{M_l}{\alpha}. \tag{5}$$

Example 1. Let Ω be a domain, $f \in L^2(\Omega)$. With $V = H_0^1(\Omega)$ we can write instead of (3):

Find $u \in H_0^1(\Omega)$ such that

$$a(u, v) := \int_\Omega \nabla u \nabla v \, dx = \int_\Omega f v \, dx =: l(v) \qquad \forall v \in H_0^1(\Omega). \tag{6}$$

Using the Cauchy–Schwarz inequality we obtain

$$a(u, v) = \int_\Omega \nabla u \nabla v \, dx \leq \|\nabla u\|_{L^2(\Omega)} \|\nabla v\|_{L^2(\Omega)} \leq \|u\|_{H^1(\Omega)} \|v\|_{H^1(\Omega)}$$

$$l(v) = \int_\Omega f v \, dx \leq \|f\|_{L^2(\Omega)} \|v\|_{L^2(\Omega)} \leq \|f\|_{L^2(\Omega)} \|v\|_{H^1(\Omega)}.$$

Using the Poincaré–Friedrichs inequality, Theorem 5, we obtain

$$a(u, u) = \int_\Omega \nabla u \nabla u \, dx = \|\nabla u\|_{L^2(\Omega)}^2 \geq \frac{1}{1 + C} \|u\|_{H^1(\Omega)}^2.$$

Therefore $a(\cdot, \cdot) : H_0^1(\Omega) \times H^1(\Omega) \to \mathbb{R}$ is continuous and elliptic and $l(\cdot) : H_0^1(\Omega) \to \mathbb{R}$ is continuous, i.e., we can apply Theorem 1, guaranteeing the existence of a unique solution of the Poisson problem.

2.3.   *Galerkin scheme and a priori estimates*

Let $V_h \subset V$ with V_h being a subspace. The Galerkin problem reads:

Find $u_h \in V_h$ such that

$$a(u_h, v_h) = l(v_h) \quad \forall v_h \in V_h. \tag{7}$$

Because V_h is a subspace of V we can apply the Lax–Milgram theorem, Theorem 1, to the Galerkin problem which proves that there exists a unique solution $u_h \in V_h$ of (7).

2.3.1. *Galerkin orthogonality*

Taking $v = v_h$ in (4) and subtracting (7) gives

$$a(u - u_h, v_h) = 0 \quad \forall v_h \in V_h. \tag{8}$$

Lemma 1 (Best approximation). *Let u be the solution of (4) and u_h be the Galerkin solution of (7). Then there holds*

$$\|u - u_h\|_V \leq \sqrt{\frac{M}{\alpha}} \|u - v_h\|_V \quad \forall v_h \in V_h. \tag{9}$$

Proof. Due to Galerkin orthogonality (8) we have for all $v_h \in V_h$

$$a(u - u_h, u - u_h) = a(u - u_h, u - v_h + v_h - u_h) = a(u - u_h, u - v_h).$$

Using ellipticity and continuity of $a(\cdot, \cdot)$ we obtain

$$\alpha \|u - u_h\|_V^2 \leq a(u - u_h, u - u_h) = a(u - u_h, u - v_h) \leq M \|u - u_h\|_V \|u - v_h\|_V.$$

$$\square$$

Example 2. Let ω_h be a quasi-uniform, regular triangulation of Ω, c.f. Section 4, i.e., $\Omega = \bigcup_{T \in \omega_h} T$, where h indicates the mesh size.

Let $V_h \subset H_0^1(\Omega)$ be the space of piecewise linear, continuous functions, subordinated to the mesh ω_h.

Together with an interpolation result we can derive an *a priori* error estimate for the Poisson problem (1). If we assume that the solution u of (6) has higher regularity, i.e., $u \in H^2(\Omega)$, the Sobolev embedding theorem 4 guarantees $u \in C^0(\bar{\Omega})$. Therefore, the nodal interpolation $\Pi_h u \in V_h$ is uniquely defined by the interpolation property

$$(\Pi_h u)(x) = u(x) \quad \text{for all mesh nodes } x.$$

Then the approximation theorem [2, Theorem 4.4.20] reads, c.f. Theorem 6:

$$\|u - \Pi_h u\|_{H^1(\Omega)} \leq C h \|u\|_{H^2(\Omega)},$$

with C independent of h and u.

Using the best approximation result, Lemma 1, we can now conclude

$$\|u - u_h\|_{H^1(\Omega)} \leq \sqrt{\frac{M}{\alpha}} \|u - \Pi_h u\|_{H^1(\Omega)} \leq C \sqrt{\frac{M}{\alpha}} h \|u\|_{H^2(\Omega)}.$$

This means, that the Galerkin method converges linearly with respect to the mesh size h, if the solution is in $H^2(\Omega)$.

Theorem 2 (Error in energy norm). *Let $a(\cdot, \cdot)$ be an elliptic and symmetric bilinear form. Let u be the solution of (4) and u_h be the Galerkin solution of (7). Then there holds*

$$||u - u_h||_a^2 = ||u||_a^2 - ||u_h||_a^2, \tag{10}$$

using the energy norm

$$||v||_a := \sqrt{a(v,v)} \quad \forall v \in V. \tag{11}$$

Remark 1. We can compute the norm of the difference $u - u_h$ without computing the difference $u - u_h$ first!

Proof. Using Galerkin orthogonality (8) there holds

$$a(u - u_h, u - u_h) = a(u, u) + a(u_h, u_h) - 2a(u, u_h) + 2\underbrace{a(u - u_h, u_h)}_{=0}$$

$$= a(u, u) + a(u_h, u_h) - 2u(u_h, u_h)$$

$$= a(u, u) - a(u_h, u_h). \qquad \square$$

2.4. *Discretisation*

In the following we will rewrite (7) in form of a linear system.

Let $\phi_1, \ldots, \phi_N$ be a basis of V_h, i.e., $\phi_i \in V_h$ for $i = 1, \ldots, N$; $\phi_1, \ldots, \phi_N$ are linear independent and for every $v_h \in V_h$ we have $d = (d_1, \ldots, d_N) \in \mathbb{R}^N$ such that $v_h = \sum_{i=1}^N d_i \phi_i$. With other words

$$V_h = \text{span}\{\phi_1, \ldots, \phi_N\}.$$

Due to $\phi_i \in V_h$ for all $i = 1, \ldots, N$ and $u_h = \sum_{j=1}^N c_j \phi_j$ the Galerkin formulation gives

$$\sum_{j=1}^N c_j \underbrace{a(\phi_j, \phi_i)}_{=:a_{ij}} = a(u_h, \phi_i) = \underbrace{l(\phi_i)}_{=:b_i}, \quad i = 1, \ldots, N. \tag{12}$$

Therefore, we obtain the linear system $A \cdot c = b$ with $A = (a_{ij})_{i,j=1,\ldots,N} \in \mathbb{R}^{N \times N}$ and $b = (b_i)_{i=1,\ldots,N} \in \mathbb{R}^N$.

The Galerkin scheme in matrix form now reads:

Find $c = (c_1, \ldots, c_N) \in \mathbb{R}^N$ such that

$$Ac = b. \tag{13}$$

Theorem 3. *Problems* (7) *and* (13) *are equivalent, i.e., if* $u_h = \sum_{j=1}^{N} c_j \phi_j$
is the solution of (7) *then* $c = (c_1, \ldots, c_N)$ *is the solution of* (13); *if* $c = (c_1, \ldots, c_N)$ *is the solution of* (13), *then the solution of* (7) *is given by*

$$u_h = \sum_{j=1}^{N} c_j \phi_j.$$

Proof. From the construction of problem (13) it follows that any solution of (7) is also a solution of (13).

Let $u_h = \sum_{j=1}^{N} c_j \phi_j$ be now the solution of (13) and $v_h = \sum_{i=1}^{N} d_i \phi_i \in V_h$ a test function. Then it follows from (12) using the linearity of $a(\cdot, \cdot)$ and $l(\cdot)$ that we have

$$a(u_h, v_h) = \sum_{i=1}^{N} d_i \sum_{j=1}^{N} c_j a(\phi_j, \phi_i) = \sum_{i=1}^{N} d_i l(\phi_i) = l(v_h).$$

$\square$

Remark 2. We do not need to compute u_h to compute the energy norm $\|u_h\|_a$. There holds

$$\|u_h\|_a^2 = a(u_h, u_h) = a\left(\sum_{i=1}^{N} c_i \phi_i, \sum_{j=1}^{N} c_j \phi_j\right) = \sum_{i=1}^{N} \sum_{j=1}^{N} c_i c_j a_{ij} = c^t A c.$$

If $u_h \in V_h$ is the solution of (13) then we even have

$$\|u_h\|_a^2 = c^t A c = c^t b.$$

The properties of the bilinear form $a(\cdot, \cdot)$ are reflected in the matrix A.

Lemma 2. (i) *Let the bilinear form* $a(\cdot, \cdot)$ *be symmetric, then* A *is symmetric.*

(ii) *Let the bilinear form* $a(\cdot, \cdot)$ *be elliptic, then* A *is positive definite.*

Proof. (i) There holds

$$a_{ij} = a(\phi_j, \phi_i) = a(\phi_i, \phi_j) = a_{ji}.$$

(ii) There holds for $u_h = \sum_{i=1}^{N} c_i \phi_i$

$$c^t A c = \sum_{i=1}^{N} \sum_{j=1}^{N} c_i c_j a_{ij} = a(u_h, u_h) \geq \alpha \|u_h\|_V^2 \geq 0$$

and $\|u_h\|_V = 0$ only if $u_h = 0$, i.e., only if $c = 0$. Therefore the matrix A is positive definite.

$\square$

2.5. Postprocessing

2.5.1. Extrapolation

Usually we do not know the solution u or the value $||u||_a$. We can only compute $u_h \in V_h$ for a sequence of meshes ω_{h_i}, $i = 1, \ldots, n$. As a consequence we can compute a sequence of energy norms

$$\gamma_i := ||u_{h_i}||_a^2, \quad i = 1, \ldots, n.$$

Using a convergence acceleration scheme, like Aitken's extrapolation technique [4], we can construct a sequence which is much faster convergent than γ_i. From this new sequence we can guess a very good estimate for $||u||_a^2$.

Aitken's extrapolation technique reads:

$$\delta_i := \frac{\gamma_i \gamma_{i-2} - \gamma_{i-1}^2}{\gamma_i - 2\gamma_{i-1} + \gamma_{i-2}}, \quad i = 3, \ldots, n.$$

2.5.2. Numerical convergence rate

Defining the error $e_i = ||u - u_{h_i}||_a$ (or any similar error), we now have a sequence of pairs: $(h_1, e_1), (h_2, e_2), (h_3, e_3), \ldots$ or with the number of degrees of freedom: $(N_1, e_1), (N_2, e_2), (N_3, e_3), \ldots$. By assuming that the error satisfies an algebraic convergence law, as indicated by the *a priori* error estimate,

$$||u - u_h||_a \leq C \cdot N^\alpha$$

and assuming that the error estimate is asymptotically sharp, we can write

$$e_i = C \cdot N_i^\alpha.$$

We can compute now the numerical convergence rate α by

$$\alpha_i = \frac{\log(e_i/e_{i-1})}{\log(N_i/N_{i-1})}. \tag{14}$$

3. Interpolation results

In this section, we will investigate how functions can be approximated depending on the mesh and the regularity of the function. We will investigate the h-version on a uniform mesh, see Theorem 6, the h-version on a graded mesh, see Theorem 7 and the p-version on a uniform mesh, see Theorem 8. All results can be generalised to higher dimensions, see e.g., Refs. [5] and [6].

The typical Sobolev spaces in one dimension are given by Definition 2. The regularity of the model function in Example 3 depends on the parameter α. For a better description of this dependency we define weighted Sobolev spaces, see Definition 3.

Definition 2 (Sobolev spaces in one dimension). Let $I \subset \mathbb{R}$ be an interval. In one dimension then we have the Sobolev norms and spaces

$$\|u\|^2_{L^2(I)} = \int_I (u(x))^2 \, dx, \qquad L^2(I) := \{u \,|\, \|u\|_{L^2(I)} < \infty\},$$

$$\|u\|^2_{H^1(I)} = \|u\|^2_{L^2(I)} + \|u'\|^2_{L^2(I)}, \qquad H^1(I) := \{u \,|\, \|u\|_{H^1(I)} < \infty\},$$

$$\|u\|^2_{H^2(I)} = \|u\|^2_{H^1(I)} + \|u''\|^2_{L^2(I)}, \qquad H^2(I) := \{u \,|\, \|u\|_{H^2(I)} < \infty\}.$$

Example 3. We can compute for $u(x) := x^\alpha$

$$u(x) = x^\alpha, \qquad \|u\|^2_{L^2([0,1])} = \frac{1}{2\alpha + 1}, \quad \alpha > -1/2;$$

$$u'(x) = \alpha x^{\alpha-1}, \qquad \|u'\|^2_{L^2([0,1])} = \frac{\alpha^2}{2\alpha - 1}, \quad \alpha > 1/2;$$

$$u''(x) = \alpha(\alpha-1)x^{\alpha-2}, \qquad \|u''\|^2_{L^2([0,1])} = \frac{\alpha^2(\alpha-1)^2}{2\alpha - 3}, \quad \alpha > 3/2.$$

That is we have $u \in L^2([0,1])$ for $\alpha > -1/2$, $u \in H^1([0,1])$ for $\alpha > 1/2$ and $u \in H^2([0,1])$ for $\alpha > 3/2$.

We observe that $u(x) = x^\alpha$ is in the $H^1([0,1])$-space for $\alpha > 1/2$ but not in $H^2([0,1])$, unless α is also greater than $3/2$. There is a way to describe this apparent gap using weighted Sobolev spaces, and we will see in Theorem 7 that the loss in convergence rate can be compensated by using a graded mesh.

Definition 3 (Weighted Sobolev spaces in one dimension). Let $I = [0,1]$ be an interval. For $0 \le \beta < 1$, $k \ge 1$ the weighted Sobolev space $H^k_\beta([0,1])$ is defined by

$$\|u\|^2_{H^k_\beta(I)} := \|u\|^2_{H^{k-1}(I)} + \int_I (x^\beta u^{(k)}(x))^2 \, dx,$$

$$H^k_\beta(I) := \{u \,|\, \|u\|_{H^k_\beta(I)} < \infty\}.$$

Example 4. For $u(x) = x^\alpha$, $1/2 < \alpha \le 3/2$ we obtain

$$\int_0^1 (x^\beta u''(x))^2 \, dx = \frac{\alpha^2(\alpha-1)^2}{2\beta + 2\alpha - 3}.$$

Consequently the integral exists for $\alpha + \beta > 3/2$. For $1/2 < \alpha \leq 3/2$ we can always find a value $\beta > 3/2 - \alpha$, so that $\alpha + \beta > 3/2$, i.e., we have $u \in H_\beta^2([0,1])$ for $1 > \beta > 3/2 - \alpha$.

3.1. *Main interpolation theorems*

In the following, we will use nodal interpolation operators, i.e., our functions have to be continuous to allow point evaluation. The following general Theorem 4 and the special case for one dimension, Lemma 3, ensure that a high-order Sobolev space implies continuity.

Theorem 4 (Sobolev embedding theorem). *[7] Let $\Omega \subset \mathbb{R}^d$ be open, Lipschitz and bounded. For*

$$m - \frac{d}{2} = k + \alpha, \quad 0 < \alpha < 1,$$

holds

$$H^m(\Omega) \subset C^{k,\alpha}(\bar{\Omega}).$$

Lemma 3. *Let $u \in H^1([0,1])$. There holds for all $x, y \in [0,1]$*

$$|u(x) - u(y)| \leq |x - y|^{0.5} \|u\|_{H^1([0,1])} \quad \Longrightarrow \quad H^1([0,1]) \subset C^{0,1/2}([0,1]).$$

Proof. Using the Cauchy–Schwarz inequality we obtain

$$|u(x) - u(y)|^2 = |\int_x^y u'(z)\,dz|^2 \leq |x - y| \cdot |\int_x^y (u'(z))^2\,dz|$$

$$\leq |x - y| \cdot \int_0^1 (u'(z))^2\,dz.$$

$\square$

Theorem 5 (Poincaré–Friedrichs inequality). *[2, Equation (5.3.3)] Let $\Omega \subset \mathbb{R}^d$ be a bounded star-shaped domain, then there holds*

$$\|v\|_{L^2(\Omega)} \leq C \left(|\int_{\partial\Omega} v\,ds| + \|\nabla v\|_{L^2(\Omega)} \right) \quad \forall v \in H^1(\Omega),$$

where the constant $C > 0$ only depends on Ω and $\partial\Omega$.

Lemma 4 (Poincaré–Friedrichs inequality in one dimension). *There holds*

$$\int_0^1 v^2(x)\,dx \leq 2v^2(0) + \int_0^1 (v'(y))^2\,dy \quad \forall v \in H^1([0,1]).$$

Proof. We apply the Cauchy–Schwarz inequality

$$\int_0^1 (v(x))^2\, dx = \int_0^1 (v(0) + v(x) - v(0))^2\, dx$$

$$= \int_0^1 \left(v(0) + \int_0^x v'(y)\, dy\right)^2 dx$$

$$\leq 2 \int_0^1 \left(v^2(0) + \left(\int_0^x v'(y)\, dy\right)^2\right) dx$$

$$\leq 2v^2(0) + 2 \int_0^1 x \int_0^x (v'(y))^2\, dy\, dx$$

$$\leq 2v^2(0) + \int_0^1 (v'(y))^2\, dy.$$

$\square$

Theorem 6. *Let* $I = [0,1]$ *and let* $u \in H^2(I)$. *Let* $x_0, x_1, \ldots, x_n$ *be a uniform mesh on* $[0,1]$, *i.e.,* $x_i = i \cdot h$, $i = 0, \ldots, n$ *with* $h = 1/n$. *Let* $u_h \in S_h^1 := \{v \in C^0([0,1]) \,|\, v|_{[x_{i-1}, x_i]} \in \mathbb{P}^1([x_{i-1}, x_i]), \quad i = 1, \ldots, n\}$ *with* $u_h(x_i) = u(x_i)$, $i = 0, \ldots, n$. *Then there holds*

$$\|u - u_h\|_{H^1([0,1])} \leq C\, h \|u\|_{H^2([0,1])}. \tag{15}$$

Proof. Using Corollary 2 for the linear interpolation operator we have

$$u_h|_{[x_{i-1}, x_i]} := \Pi^1_{[x_{i-1}, x_i]} u.$$

Using Lemma 5, we obtain

$$\|u' - u_h'\|_{L^2(I)}^2 = \sum_{i=1}^n \int_{x_{i-1}}^{x_i} (u' - u_h')^2\, dx \leq \sum_{i=1}^n h^2 \int_{x_{i-1}}^{x_i} (u'')^2\, dx = h^2 \|u''\|_{L^2(I)}^2.$$

Due to the interpolation condition we have in particular $u(0) - u_h(0) = 0$, i.e., we can apply the Poincaré–Friedrichs inequality Lemma 4 and we have

$$\|u - u_h\|_{H^1(I)}^2 = \|u - u_h\|_{L^2(I)}^2 + \|u' - u_h'\|_{L^2(I)}^2 \leq C\, h^2 \|u''\|_{L^2(I)}^2. \quad \square$$

Theorem 7. *Let* $\in H^2_\beta([0,1])$ *with* $0 \leq \beta < 1$. *Let* $\eta \geq 1$. *We choose the nodes*

$$x_i = \left(\frac{i}{n}\right)^\eta, \qquad i = 0, \ldots, n.$$

Let $u_h \in S^1_{n,\eta} := \{v \in C^0([0,1]) \,|\, v|_{[x_{i-1}, x_i]} \in \mathbb{P}^1([x_{i-1}, x_i]), i = 1, \ldots, n\}$ *with* $u_h(x_i) = u(x_i)$, $i = 0, \ldots, n$. *Then there exists a constant* $C > 0$

independent of $h = 1/n$, but depending on η, β, such that

$$\|u - u_h\|_{H^1([0,1])} \leq C h^\alpha \|u\|_{H^2_\beta([0,1])}, \quad \alpha = \min(1, (1 - \beta)\eta). \tag{16}$$

Proof. Due to $u \in H^2_\beta([0,1]) \subset H^1([0,1]) \subset C^0([0,1])$ the linear interpolant u_h of u is well-defined.

For the mesh width $h_k = x_k - x_{k-1}$ we have with $h = 1/n, \gamma = 1 - \frac{1}{\eta}$

$$h_k = x_k - x_{k-1} \leq \eta h x_k^{(\eta-1)/\eta} = \eta h x_k^\gamma. \tag{17}$$

Due to Lemma 6 and Corollary 3 we have

$$\|u' - u_h'\|^2_{L^2([x_{k-1},x_k])} \leq \frac{h_k^2 x_k^{-2\beta}}{1 - \beta} \|x^\beta u''\|^2_{L^2([x_{k-1},x_k])}. \tag{18}$$

Together with (17) we obtain

$$\|u' - u_h'\|^2_{L^2([0,1])} = \sum_{k=1}^n \|u' - u_h'\|^2_{L^2([x_{k-1},x_k])}$$

$$\leq \frac{1}{1 - \beta} \sum_{k=1}^n h_k^2 x_k^{-2\beta} \|x^\beta u''(x)\|^2_{L^2([x_{k-1},x_k])}$$

$$\leq \frac{1}{1 - \beta} \max_{1 \leq k \leq n} \left(h_k^2 x_k^{-2\beta} \right) \|u\|^2_{H^2_\beta([0,1])}$$

$$\leq \frac{\eta^2 h^2}{1 - \beta} \max_{1 \leq k \leq n} \left(x_k^{2(\gamma-\beta)} \right) \|u\|^2_{H^2_\beta([0,1])}.$$

We have

$$\gamma - \beta = 1 - \beta - \frac{1}{\eta} = ((1 - \beta)\eta - 1)/\eta.$$

For $(1 - \beta)\eta \geq 1$ it follows $\gamma - \beta \geq 0$ and therefore $x_k^{2(\gamma-\beta)} \leq 1$ due to $0 \leq x_{k-1} \leq 1$ and we have $\max_{1 \leq k \leq n}(x_k^{2(\gamma-\beta)}) = 1$.

For $(1 - \beta)\eta \leq 1$ we have $\gamma - \beta \leq 0$ and therefore $x_k^{2(\gamma-\beta)} \leq x_1^{2(\gamma-\beta)} = h^{2(\gamma-\beta)\eta}$ and we have $\max_{1 \leq k \leq n}(x_k^{2(\gamma-\beta)}) = h^{2(\gamma-\beta)\eta} = h^{2(1-\frac{1}{\eta}-\beta)\eta} = h^{-2+2(1-\beta)\eta}$.

Combining both results we obtain

$$\|u' - u_h'\|^2_{L^2([0,1])} \leq \frac{\eta^2}{1 - \beta} h^{2\min(1,(1-\beta)\eta)} \|u\|^2_{H^2_\beta([0,1])}.$$

The H^1-estimate follows again using the Poincaré–Friedrichs inequality, see Lemma 4. □

Theorem 8. *Let* $I = [0,1]$ *and let* $u \in H^2(I)$. *Let* $x_0, x_1, \ldots, x_n$ *be a uniform mesh on* $[0,1]$, *i.e.,* $x_i = i \cdot h$, $i = 0, \ldots, n$ *with* $h = 1/n$. *Let* $p \geq 1$ *be the polynomial degree. Then there exists* $u_{hp} \in S_h^p := \{v \in C^0([0,1]) \,|\, v|_{[x_{i-1},x_i]} \in \mathbb{P}^p([x_{i-1},x_i]), i = 1, \ldots, n\}$ *with* $u_{hp}(x_i) = u(x_i)$, $i = 0, \ldots, n$, *such that there holds*

$$\|u - u_{hp}\|_{H^1([0,1])} \leq C \frac{h}{p} \|u\|_{H^2([0,1])}. \tag{19}$$

Proof. Using Definition 4 for the interpolation operator we define

$$u_{hp}|_{[x_{i-1},x_i]} := \Pi^p_{[x_{i-1},x_i]} u.$$

In Lemma 7, we choose $k = p - 1$, $s = 0$ and we map the interval $[-1,1]$ onto $I_i := [x_{i-1}, x_i]$ using a linear transformation. Then we obtain

$$\|u' - P_{I_i}^{p-1} u'\|^2_{L^2(I_i)} \leq \frac{(p-1-s)!}{(p-1+s+2)!} \left(\frac{h}{2}\right)^{2(s+1)} \left\|\frac{\partial^{s+1} u'}{\partial x^{s+1}}\right\|^2_{L^2(I_i)}$$

$$= \frac{(p-1)!}{(p+1)!} \left(\frac{h}{2}\right)^2 \|u''\|^2_{L^2(I_i)} \leq \left(\frac{h}{p}\right)^2 \|u''\|^2_{L^2(I_i)}.$$

That is together we have

$$\|u' - u'_{hp}\|^2_{L^2(I)} = \sum_{i=1}^n \|u' - u'_{hp}\|^2_{L^2(I_i)}$$

$$\leq \sum_{i=1}^n \left(\frac{h}{p}\right)^2 \|u''\|^2_{L^2(I_i)} = \left(\frac{h}{p}\right)^2 \|u''\|^2_{L^2(I)}.$$

Due to the interpolation condition, see Corollary 1, in particular $u(0) - u_{hp}(0) = 0$, we can apply the Poincaré–Friedrichs inequality Lemma 4 and we obtain the assertion of the theorem. $\square$

3.2. *Local interpolation results*

Definition 4. On the interval $[a,b]$ we define the L^2-projection operator $P_{[a,b]}^p : L^2([a,b]) \to \mathbb{P}^p_{[a,b]} := \mathrm{span}\{t^0, t^1 \ldots, t^p\}$ by

$$\left(u - P_{[a,b]}^p u, v\right)_{L^2([a,b])} = 0 \qquad \forall v \in \mathbb{P}^p_{[a,b]}. \tag{20}$$

Existence and uniqueness of $P_{[a,b]}^p u \in \mathbb{P}^p_{[a,b]}$ follows by Theorem 1.

On the interval $[a,b]$ we define the interpolation operator $\Pi^p_{[a,b]}$: $H^1([a,b]) \to \mathbb{P}^p_{[a,b]}$ by

$$\left(\Pi^p_{[a,b]} u\right)(x) := u(a) + \int_a^x \left(P^{p-1}_{[a,b]} u'\right)(y)\, dy. \qquad (21)$$

Corollary 1. *There holds*

$$\left(\Pi^p_{[a,b]} u\right)(a) = u(a), \qquad \left(\Pi^p_{[a,b]} u\right)(b) = u(b).$$

Proof. Due to Equation (20) we have

$$\int_a^b u'(x) \cdot 1\, dx = \int_a^b \left(P^{p-1}_{[a,b]} u'\right)(x) \cdot 1\, dx,$$

i.e., we have

$$\left(\Pi^p_{[a,b]} u\right)(b) = u(a) + \int_a^b \left(P^{p-1}_{[a,b]} u'\right)(y)\, dy = u(a) + \int_a^b u'(y)\, dx$$

$$= u(a) + (u(b) - u(a)) = u(b). \qquad \square$$

Corollary 2. *In the lowest-order case there holds*

$$P^0_{[a,b]} u := \frac{1}{b-a} \int_a^b u(y)\, dy$$

and

$$\left(\Pi^1_{[a,b]} u\right)(x) = u(a) + \int_a^x \left(P^0_{[a,b]} u'\right)(y)\, dy$$

$$= u(a) + \int_a^x \frac{1}{b-a} \int_a^b u'(z)\, dz\, dy$$

$$= u(a) + \frac{x-a}{b-a}(u(b) - u(a))$$

$$= u(a)\frac{b-x}{b-a} + u(b)\frac{x-a}{b-a}.$$

Remark 3. There holds the relation

$$\left(\Pi^p_{[a,b]} u\right)'(x) = \left(P^{p-1}_{[a,b]} u'\right)(x).$$

Lemma 5 (one-dimensional-interpolation). *Let $u \in H^1([0,h])$. We will show that for the projection operator $P^0_{[0,h]}$ holds*

$$\|u - P^0_{[0,h]} u\|_{L^2([0,h])} \leq Ch\|u'\|_{L^2([0,h])}.$$

Proof. Due to Corollary 2 we have

$$\int_0^h \left(u(x) - (P^0_{[0,h]}u)(x) \right)^2 dx = \int_0^h \left(u(x) - \frac{1}{h}\int_0^h u(y)\,dy \right)^2 dx$$

$$= \frac{1}{h^2}\int_0^h \left(\int_0^h (u(x) - u(y))\,dy \right)^2 dx$$

$$= \frac{1}{h^2}\int_0^h \left(\int_0^h \left(\int_y^x u'(z)\,dz \right) dy \right)^2 dx$$

$$\leq \frac{1}{h^2}\int_0^h h \int_0^h \left(\int_y^x u'(z)\,dz \right)^2 dy\,dx$$

$$\leq \frac{1}{h^2}\int_0^h h \int_0^h |x-y| \cdot | \int_y^x (u'(z))^2\,dz|\,dy\,dx$$

$$\leq \frac{1}{h^2}\int_0^h h \int_0^h h \int_0^h (u'(z))^2\,dz\,dy\,dx$$

$$= h^2 \int_0^h (u'(z))^2\,dz.$$

$\square$

Lemma 6. *Let* $I = [a,b] \subseteq [0,1]$, $h = b - a \geq 0$ *and* $0 \leq \beta < 1$. *Let* $u \in H^1_\beta([0,1])$. *Then there holds for the* L^2-*projection* $P^0_{[a,b]}u$

$$\left\| u - P^0_{[a,b]}u \right\|^2_{L^2(I)} \leq 2 \left(\int_a^b \int_y^b z^{-2\beta}\,dz\,dy \right) \|x^\beta \partial_x u(x)\|^2_{L^2(I)}. \qquad (22)$$

Proof. The proof is a generalisation of the proof of Lemma 5

$$\int_a^b \left(u(x) - \left(P^0_{[a,b]}u \right)(x) \right)^2 dx = \int_a^b \left(u(x) - \frac{1}{h}\int_a^b u(y)\,dy \right)^2 dx$$

$$= \frac{1}{h^2}\int_a^b \left(\int_a^b (u(x) - u(y))\,dy \right)^2 dx$$

using the Cauchy–Schwarz inequality we have

$$\leq \frac{1}{h^2} \int_a^b h \int_a^b (u(x) - u(y))^2 \, dy \, dx$$

$$= \frac{1}{h} \int_a^b \int_a^b (u(x) - u(y))^2 \, dy \, dx$$

due to the symmetry of $(u(x) - u(y))^2$ we can write

$$= \frac{2}{h} \int_a^b \int_a^x (u(x) - u(y))^2 \, dy \, dx = \frac{2}{h} \int_a^b \int_a^x \left(\int_y^x \partial_z u(z) \, dz \right)^2 \, dy \, dx$$

$$= \frac{2}{h} \int_a^b \int_a^x \left(\int_y^x z^{-\beta} z^\beta \partial_z u(z) \, dz \right)^2 \, dy \, dx$$

using the Cauchy–Schwarz inequality again and the fact that $y \leq x$, we have

$$\leq \frac{2}{h} \int_a^b \int_a^x \int_y^x z^{-2\beta} \int_y^x z^{2\beta} (\partial_z u(z))^2 \, dz \, dy \, dx$$

$$\leq 2 \left(\int_a^b \int_y^b z^{-2\beta} \, dz \, dy \right) \int_a^b z^{2\beta} (\partial_z u(z))^2 \, dz. \qquad \square$$

Corollary 3. *For $0 \leq a \leq b \leq 1$ and $0 \leq \beta < 1$ there holds*

$$\int_a^b \int_y^b z^{-2\beta} \, dz \, dy \leq \frac{(b-a)^2 b^{-2\beta}}{2(1-\beta)}. \qquad (23)$$

Proof. First we normalise the integral using the substitution $y = b \cdot u$, $z = b \cdot v$

$$\int_a^b \int_y^b z^{-2\beta} \, dz \, dy = b^{2-2\beta} \int_{a/b}^1 \int_u^1 v^{-2\beta} \, dv \, du =: b^{2-2\beta} I_{a/b}.$$

For $0 \leq \beta < 1$, $\beta \neq 1/2$ the integral I_c ($c = a/b \in [0,1]$) has the value

$$I_c = \int_c^1 \int_u^1 v^{-2\beta} \, dv \, du = \int_c^1 \left[\frac{v^{1-2\beta}}{1-2\beta} \right]_u^1 \, du = \frac{1}{1-2\beta} \int_c^1 (1 - u^{1-2\beta}) \, du$$

$$= \frac{1}{1-2\beta} \left((1-c) - \left[\frac{u^{2-2\beta}}{2-2\beta} \right]_c^1 \right) = \frac{1-c}{1-2\beta} - \frac{1 - c^{2-2\beta}}{(1-2\beta)(2-2\beta)}$$

$$= \frac{(1-c)^2}{2-2\beta} + \frac{c}{2-2\beta} \left((1-c) - \frac{1 - c^{1-2\beta}}{1-2\beta} \right).$$

We define $f(c) := (1-c) - \frac{1-c^{1-2\beta}}{1-2\beta}$ and show that $f(c) \leq 0$, $c \in [0,1]$ due to

$$f(1) = 1 - 1 - \frac{1 - 1^{1-2\beta}}{1 - 2\beta} = 0,$$

$$f'(c) = -1 - \frac{-(1-2\beta)c^{-2\beta}}{1-2\beta} = -1 + c^{-2\beta} \geq 0, \quad c \in [0,1].$$

Consequently, we have

$$\int_a^b \int_y^b z^{-2\beta} \, dz \, dy = b^{2-2\beta} I_{a/b} \leq b^{2-2\beta} \frac{(1-a/b)^2}{2-2\beta} = \frac{(b-a)^2 b^{-2\beta}}{2-2\beta}.$$

For $\beta = 1/2$ we can compute more directly

$$\int_a^b \int_y^b z^{-1} \, dz \, dy = \int_a^b [\log z]_y^b \, dy = \int_a^b (\log b - \log y) \, dy$$

$$= (b-a)\log b - [y \log y - y]_a^b$$

$$= (b-a)\log b - b \log b + b + a \log a - a$$

$$= -a \log b + a \log b + b - a$$

$$= a \log \left(1 + \left(-1 + \frac{a}{b}\right)\right) + b - a$$

$$\leq a \left(-1 + \frac{a}{b}\right) + b - a = -a + \frac{a^2}{b} + b - a = \frac{(b-a)^2}{b}.$$

$\square$

Definition 5 (Legendre-polynomials). The Legendre polynomials $L_n(x)$, $x \in \mathbb{R}$ are defined by

$$L_n(x) = \frac{1}{2^n \, n!} \frac{d^n}{dx^n} (x^2 - 1)^n, \quad n \geq 0. \tag{24}$$

Remark 4. For any value $x \in \mathbb{R}$ the vector of $(p+1)$ function values $(L_n(x))_{n=0,\ldots,p}$ can be computed efficiently with complexity $\mathcal{O}(p)$ using the recurrence formula

$$nL_n(x) = (2n-1)xL_{n-1}(x) - (n-1)L_{n-2}(x), n = 2, \ldots, p, \tag{25}$$

with $L_0(x) = 1$, $L_1(x) = x$.

Lemma 7 (cf. [8]). *Let $I = (-1,1)$ be an interval and $u \in H^{s+1}(I)$, integers $k \geq s \geq 0$, then there exists a polynomial $\phi := P^k_{[-1,1]}u \in \mathbb{P}^k_{[-1,1]}$, such that there holds*

$$\|u - P^k_{[-1,1]}u\|^2_{L^2(I)} \leq \frac{(k-s)!}{(k+s+2)!} \left\| \frac{\partial^{s+1}u}{\partial x^{s+1}} \right\|^2_{L^2(I)}. \tag{26}$$

Proof. Due to $u \in H^1(I) \subset L^2(I)$ and the fact that the Legendre polynomials are a orthogonal basis in $L^2(I)$ we can write

$$u(x) = \sum_{i=0}^{\infty} c_i L_i(x), \quad x \in (-1, 1).$$

We have $\phi := P_{[-1,1]}^k u$, i.e., ϕ is the L^2 projection of u on $\mathbb{P}_{[-1,1]}^k$; due to the orthogonality of the Legendre polynomials $\phi(x)$ is given by the first $k+1$ terms in the expansion of u into Legendre polynomials and we obtain for the difference $u - \phi$

$$\phi(x) = \sum_{0 \le i \le k} c_i L_i(x), \quad u(x) - \phi(x) = \sum_{i \ge k+1} c_i L_i(x).$$

Due to the orthogonality relations of the Legendre polynomials we have

$$\|u - \phi\|_{L^2(I)}^2 = \sum_{i \ge k+1} |c_i|^2 \frac{2}{2i+1} = \frac{(k-s)!}{(k+s+2)!}$$

$$\sum_{i \ge k+1} |c_i|^2 \frac{2}{2i+1} \frac{(k+s+2)!}{(k-s)!}.$$

For $i \ge k+1$ we have

$$\frac{(k+s+2)!}{(k-s)!} \le \frac{(i+s+1)!}{(i-s-1)!}$$

and we obtain

$$\|u - \phi\|_{L^2(I)}^2 \le \frac{(k-s)!}{(k+s+2)!} \sum_{i \ge k+1} |c_i|^2 \frac{2}{2i+1} \frac{(i+s+1)!}{(i-s-1)!}$$

$$\le \frac{(k-s)!}{(k+s+2)!} \sum_{i \ge s+1} |c_i|^2 \frac{2}{2i+1} \frac{(i+s+1)!}{(i-s-1)!}.$$

Due to [8, Lemma 4.1] holds

$$\int_I |\partial_x^\alpha u|^2 (1-x^2)^\alpha \, dx = \sum_{i \ge \alpha} |c_i|^2 \frac{2}{2i+1} \frac{(i+\alpha)!}{(i-\alpha)!}$$

and therefore

$$\sum_{i \ge s+1} |c_i|^2 \frac{2}{2i+1} \frac{(i+s+1)!}{(i-(s+1))!} = \int_I \left|\frac{\partial^{s+1} u}{\partial x^{s+1}}\right|^2 (1-x^2)^{s+1} \, dx \le \left\|\frac{\partial^{s+1} u}{\partial x^{s+1}}\right\|_{L^2(I)}^2.$$

Altogether we have

$$\|u - \phi\|_{L^2(I)}^2 \le \frac{(k-s)!}{(k+s+2)!} \left\|\frac{\partial^{s+1} u}{\partial x^{s+1}}\right\|_{L^2(I)}^2. \qquad \square$$

4. Implementation

In this section, we introduce algorithms to set-up the linear system (13) for the Poisson problem in two dimensions with a triangular mesh $\Omega = \cup_{i=1}^{N_{\text{elem}}} \Omega_i$ and piecewise linear and continuous basis functions. The described algorithms can be easily generalised to cover more general elements, higher polynomial basis functions and different PDEs. We also describe algorithms for numerical quadrature, optimisation techniques, and the conjugate gradient solver.

Our fundamental assumptions are the following:

- Our (global, continuous) basis functions are based on a mesh.
- On every mesh element we have a given set of local basis functions.
- The local basis functions are generated by mapping a reference element to the mesh element.
- Basis functions are a linear combination of local basis functions.
- Every local basis functions belongs to one and only one global basis function.

Using this fundamental assumptions we have the following objects to deal with:

- The mesh consisting of mesh elements: Ω_i.
- The mappings $F_i : Q \mapsto \Omega_i$ from the reference element to the mesh elements.
- The set of basis function on the reference element: $\phi_k^{\text{ref}} : Q \mapsto \mathbb{R}$.
- The local basis functions on every element: $\phi_{i,k}^{\text{loc}}(x) = \phi_k^{\text{ref}}(F_i^{-1}(x)) : \Omega_i \mapsto \mathbb{R}$.

Then every global basis function can be represented in the following way:

$$\phi_p(x) = \sum_{p=\text{map}(i,k)} \phi_{i,k}^{\text{loc}}(x). \tag{27}$$

Here, $\text{map}(i,k)$ denotes the global basis functions to which every local basis function belongs to.

In case of triangles and piecewise linear basis functions, we have the following basis functions on the reference element $Q = \{(t_1, t_2) \,|\, 0 \leq t_1 \leq 1, 0 \leq t_2 \leq 1 - t_1\}$

$$\phi_1^{\text{ref}}(t_1, t_2) = 1 - t_1 - t_2;$$

Global node numbers Element numbers Local node numbers Edge numbers

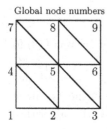 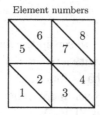 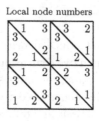 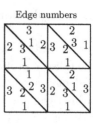

Fig. 1. Geometrical information.

$$\phi_2^{\mathrm{ref}}(t_1, t_2) = t_1;$$

$$\phi_3^{\mathrm{ref}}(t_1, t_2) = t_2.$$

We note that the choice of basis functions allows to construct a nodal basis, i.e., for the three vertices of Q, i.e., $(0,0),(1,0),(0,1)$ we have e.g., $\phi_1^{\mathrm{ref}}(0,0) = 1, \phi_1^{\mathrm{ref}}(1,0) = 0, \phi_1^{\mathrm{ref}}(0,1) = 0$ etc.

We have N, number of global basis functions, N_{elem}, number of elements and N_{nodes} the number of nodes. There holds $N \approx N_{\mathrm{elem}} \approx N_{\mathrm{nodes}}$. The mesh is defined by the unique set of nodes $x_m \in \mathbb{R}^2, m = 1, \ldots, N_{\mathrm{nodes}}$, and the nodes-table which stores for every element Ω_i and every vertex (local node number) the global node number, i.e., $m = \mathrm{nodes}(i,k)$, $i = 1, \ldots, N_{\mathrm{elem}}, k = 1, \ldots, 3$, see Fig. 1. The mapping $F_i : Q \mapsto \Omega_i$ is an affine mapping, if Ω_i is a triangle. Based on the nodes-table we can also define local edge numbers, i.e., local edge 1 is defined by local nodes 1 and 2, local edge 2 is defined by local nodes 2 and 3 and local edge 3 is defined by local nodes 3 and 1.

4.1. Assembling the right-hand side

Using the piecewise defined basis functions (27) we obtain the following expression for $b_p = \int_\Omega f(x)\phi_p(x)\,dx$, $p = 1, \ldots, N$

$$b_p = \int_\Omega f(x)\phi_p(x)\,dx = \int_\Omega f(x) \sum_{p=\mathrm{map}(i,k)} \phi_{i,k}^{\mathrm{loc}}(x)\,dx$$

$$= \sum_{p=\mathrm{map}(i,k)} \int_\Omega f(x)\phi_{i,k}^{\mathrm{loc}}(x)\,dx = \sum_{p=\mathrm{map}(i,k)} \underbrace{\int_{\Omega_i} f(x)\phi_{i,k}^{\mathrm{loc}}(x)\,dx}_{=:b_{i,k}^{\mathrm{elem}}}.$$

For the local integrals we obtain after transformation to the reference element

$$b_{i,k}^{\text{elem}} = \int_{\Omega_i} f(x)\phi_k^{\text{ref}}(F_i^{-1}(x))\,dx = \int_Q f(F_i(t))\phi_k^{\text{ref}}(t)\left|\frac{\partial F_i(t)}{\partial t}\right|\,dt.$$

If we tread the expression above literally, we would have to compute N values b_p and for every value of p we would have to compute a sum, i.e., N_{elem} terms (number of elements). Therefore, the total complexity is $\mathcal{O}(N \cdot N_{\text{elem}}) \sim \mathcal{O}(N^2)$.

But we notice that we only have $\sim N_{\text{elem}}$ local basis functions $\phi_{i,k}^{\text{loc}}$, therefore we have only to compute $\sim N_{\text{elem}}$ local integrals. The following algorithm now computes the integrals for every element first and then adds the values to the right hand side vector. The new complexity is now $\mathcal{O}(N)$.

Algorithm 1 (r.h.s. assembly). This algorithm assembles the right-hand side.

1: $b_{1,\ldots,N} \leftarrow 0$
2: **for** $i = 1,\ldots,N_{\text{elem}}$ **do**
3:     **for all** k **do**
4:         $b_{i,k}^{\text{elem}} \leftarrow \int_\Omega f(x)\phi_{i,k}^{\text{loc}}(x)\,dx$
5:     **end for**
6:     **for all** k **do**
7:         $p \leftarrow \text{map}(i,k)$
8:         **if** $p > 0$ **then**
9:             $b_p \leftarrow b_p + b_{i,k}^{\text{elem}}$
10:         **end if**
11:     **end for**
12: **end for**

4.2. *Assembling the matrix*

Using the piecewise defined basis functions (27) we obtain the following expression for $a_{pq} = \int_\Omega \nabla_x\phi_p(x)\nabla_x\phi_q(x)\,dx,\ p,q = 1,\ldots,N$

$$a_{pq} = \int_\Omega \nabla_x\phi_p\nabla_x\phi_q\,dx = \int_\Omega \nabla_x\sum_{p=\text{map}(i,k)}\phi_{i,k}^{\text{loc}}(x)\,\nabla_x\sum_{q=\text{map}(j,l)}\phi_{j,l}^{\text{loc}}(x)\,dx$$

$$= \sum_{p=\text{map}(i,k)} \sum_{q=\text{map}(j,l)} \int_{\Omega} \nabla_x \phi_{i,k}^{\text{loc}}(x) \nabla_x \phi_{j,l}^{\text{loc}}(x)\, dx$$

$$= \sum_{p=\text{map}(i,k)} \sum_{q=\text{map}(i,l)} \underbrace{\int_{\Omega_i} \nabla_x \phi_{i,k}^{\text{loc}}(x) \nabla_x \phi_{i,l}^{\text{loc}}(x)\, dx}_{=:a_{i,k,l}^{\text{elem}}}.$$

Due to the identity $t = F_i^{-1}(F_i(t))$ there holds $I = \nabla_x F_i^{-1} \cdot \nabla_t F_i$, i.e., we can define the Jacobian matrix $J_i := \nabla_t F_i(t)$. Then we obtain

$$\nabla_x \phi_{i,k}^{\text{loc}}(x) = \nabla_x \phi_k^{\text{ref}}(F_i^{-1}(x)) = (J_i^{-1})^T \nabla_t \phi_k^{\text{ref}}(F_i^{-1}(x)).$$

For the local integrals we obtain after transformation to the reference element

$$a_{i,k,l}^{\text{elem}} = \int_Q \left(J_i^{-T} \nabla_t \phi_k^{\text{ref}}(t) \right) \left(J_i^{-T} \phi_l^{\text{ref}}(t) \right) \left| \frac{\partial F_i}{\partial t} \right| dt.$$

Again the complexity is very high, naively two loops for p, q and two loops for the sums, $\mathcal{O}(N^4)$. Whereas using the mapping information for local basis functions to global basis functions, the complexity is reduced to $\mathcal{O}(N)$ in the following algorithm.

Algorithm 2 (matrix assembly). This algorithm assembles the Galerkin matrix and can make use of the sparse structure.

1: **for** $i = 1, \ldots, N_{\text{elem}}$ **do**
2:     **for all** k,l **do**
3:         $a_{i,k,l}^{\text{elem}} \leftarrow \int_{\Omega} \nabla_x \phi_{i,k}^{\text{loc}}(x) \nabla_x \phi_{i,l}^{\text{loc}}(x)\, dx$
4:     **end for**
5:     **for all** k **do**
6:         $p \leftarrow \text{map}(i, k)$
7:         **for all** l **do**
8:             $q \leftarrow \text{map}(i, l)$
9:             **if** $p > 0$ and $q > 0$ **then**
10:                 $a_{p,q} \leftarrow a_{p,q} + a_{i,k,l}^{\text{elem}}$
11:             **end if**
12:         **end for**
13:     **end for**
14: **end for**

4.3.   *Constructing the neigh-table*

For imposing the boundary conditions and for evaluating the jump terms in the error estimator, see (33), we need for every element i and every edge k the number of the neighbouring element $j = \mathrm{neigh}(i, k)$.

A naive algorithm to fill this neigh-table would be a double loop and pairwise comparison of all edges whether they coincide, leading to an $\mathcal{O}(N_{\mathrm{elem}}^2)$ algorithm. The following algorithm first constructs a table 'elem' which maps node numbers to element numbers. This operation is of order $\mathcal{O}(N_{\mathrm{elem}})$ using the nodes-table. The number of elements connected to the same node is bounded in a regular mesh. Therefore, using pairwise comparisons does not increase the complexity further.

Algorithm 3 (neigh-creation). This algorithm constructs the neigh-table in an efficient way.

```
 1: noel(1 : Nnodes) ← 0
 2: for i = 1, ..., Nelem do
 3:     for k = 1, ..., 3 do
 4:         node ← nodes(i, k)
 5:         noel(node) ← noel(node) + 1
 6:         elem(node, noel(node)) ← i
 7:     end for
 8: end for
 9: for node = 1, ..., Nnodes do
10:     for p = 1, ..., noel(node) do
11:         i ← elem(node, p)
12:         for q = p + 1, ..., noel(node) do
13:             j ← elem(node, q)
14:             for k = 1, ..., 3 do
15:                 for l = 1, ..., 3 do
16:                     if edge k of element i == edge l of element j then
17:                         neigh(i, k) ← j
18:                         neigh(j, l) ← i
19:                     end if
20:                 end for
21:             end for
22:         end for
23:     end for
24: end for
```

4.4. Constructing the map-table

The following algorithm constructs the map-table for nodal basis functions on triangles for our model problem. The nodes-table contains already unique node numbers for every vertex of a triangle, therefore the only difference to map is the fact that we have homogenous boundary conditions, i.e., we have no basis functions belonging to the boundary nodes. In the following we start with the nodes-table, eliminate all basis functions belonging to the boundary and ensure that the remaining basis functions are consecutively numbered.

Algorithm 4 (map-creation). This algorithm constructs the map-table for nodal basis functions on triangles for an example with homogenous boundary conditions. The only input-tables are nodes and neigh.

1: $nod(1 : N_{\text{nodes}}) \leftarrow 1$
2: **for** $i = 1, \ldots, N_{\text{elem}}$ **do**
3:    $n(1 : 3) \leftarrow nodes(i, 1 : 3); n(4) \leftarrow n(1)$
4:    **for** $k = 1, \ldots, 3$ **do**
5:        **if** $\text{neigh}(i, k) == 0$ **then**
6:            $nod(n(k)) \leftarrow 0; \quad nod(n(k + 1)) \leftarrow 0$
7:        **end if**
8:    **end for**
9: **end for**
10: $cnt \leftarrow 1;$
11: **for** $i = 1, \ldots, N_{\text{nodes}}$ **do**
12:    **if** $nod(i) > 0$ **then**
13:        $nod(i) \leftarrow cnt; cnt \leftarrow cnt + 1$
14:    **end if**
15: **end for**
16: **for** $i = 1, \ldots, N_{\text{elem}}$ **do**
17:    **for** $k = 1, \ldots, 3$ **do**
18:        $\text{map}(i, k) \leftarrow nod(\text{nodes}(i, k))$
19:    **end for**
20: **end for**

4.5. Numerical quadrature

In many cases the integrals occurring in our Galerkin method cannot be computed exactly. Therefore, we have to apply a numerical quadrature scheme. There are specialised formulas on triangles and rectangles available, where the number of quadrature points is minimised. But a very simple way to construct a numerical quadrature method on triangles etc. is based on using a one dimensional formula. The Gauss–Legendre quadrature formulas have the highest accuracy for smooth functions.

For the Gauss–Legendre quadrature formulas in the interval $(-1, 1)$ we define

$$I_n f := \sum_{j=0}^{n} w_j f(x_j), \tag{28}$$

the points $x_{ni}, i = 1, \ldots, n$ are given as the roots of the Legendre polynomial $L_n(x)$ (see Definition 5), i.e.,

$$L_n(x_{ni}) = 0, \quad i = 1, \ldots, n.$$

Then the weights w_{ni} can be computed by

$$w_{ni} = \frac{2}{(1 - x_{ni}^2)(L_n'(x_{ni}))^2}.$$

The error is given by

$$\int_{-1}^{1} f(x)\, dx - I_n f = \frac{2^{2n+1}(n!)^4}{(2n+1)[(2n)!]^3} f^{(2n)}(\xi), \quad \xi \in (-1, 1).$$

On a rectangle $[-1, 1]^2$ we have now the quadrature formula

$$\int_{-1}^{1} \int_{-1}^{1} f(x, y)\, dy\, dx \approx \sum_{i=0}^{n} \sum_{j=0}^{n} w_i w_j f(x_i, x_j).$$

Using the Duffy-transformation $(t_1, t_2) := (q_1, (1 - q_1) q_2)$ which maps $[0, 1]^2$ onto the triangle $\{(t_1, t_2) : 0 \leq t_1 \leq 1, 0 \leq t_2 \leq 1 - t_1\}$, we can rewrite the integral on a triangle into an integral on a square.

$$\int_{\Delta} f := \int_{0}^{1} \int_{0}^{1-t_1} f(t_1, t_2)\, dt_2\, dt_1 = \int_{0}^{1} \int_{0}^{1} f(q_1, (1 - q_1) q_2)(1 - q_1)\, dq_2\, dq_1.$$

Using an affine transformation we can map any triangle or rectangle onto the reference elements used above.

4.6. Sum-factorisation

In the case that we are using quadrilateral elements in our mesh, so that the reference element is the square $[-1, 1]$ and we use basis functions which are the product of 1d-functions we have to compute for $r, s = 0, \ldots, p$

$$b_{rs} = \int_{-1}^{1} \int_{-1}^{1} f(x_1, x_2) \mathcal{L}_r(x_1) \mathcal{L}_s(x_2) \, dx_2 \, dx_1$$

$$\approx \sum_{m=1}^{g} \sum_{n=1}^{g} f(t_m, t_n) \mathcal{L}_r(t_m) \mathcal{L}_s(t_n) w_m w_n$$

with the g quadrature nodes $t_1, \ldots, t_g$ and weights $w_1, \ldots, w_g$. It is easy to see that the total complexity of the numerical quadrature in the form above is $\mathcal{O}((p+1)^2 g^2)$. If the number of quadrature points g is chosen as $g = p$, we obtain the complexity $\mathcal{O}(p^4)$. If we define $a_{mn} := f(t_m, t_n)$, $c_{rm} = \mathcal{L}_r(t_m) w_m$ and $d_{sm} = \mathcal{L}_s(t_n) w_n$ we have

$$b_{rs} \approx \sum_{m=1}^{g} \sum_{n=1}^{g} a_{mn} c_{rm} d_{sn}$$

or in other words, using matrix notation $A = (a_{mn})$ etc., we obtain

$$B = C \cdot A \cdot D^t.$$

Alternatively, we could also write

$$E = C \cdot A, \qquad B = E \cdot D^t,$$

therefore we have two matrix–matrix multiplication with a total complexity $\mathcal{O}(p^3)$, which is for large p considerably more efficient.

Consequently, all algorithms/implementations should be checked for hidden features, where sum-factorisation can be applied.

4.7. Solver

The Conjugate Gradient (CG)-algorithm [9, Algorithm 6.18] is an iterative solver for linear systems of the form $Ax = b$, $A \in \mathbb{R}^{N \times N}, b \in \mathbb{R}^N$, where the matrix A is symmetric and positive definite. The CG-algorithm is a Krylov subspace method, i.e., it is based on an orthogonalisation principle.

The algorithm requires on entry an initial value for x, the initial value can be just a zero-vector, and an error parameter ε. On exit the vector x is a very good approximation to the exact solution. $\langle \cdot, \cdot \rangle$ is the Euclidean inner product in $\mathbb{R}^N$.

Algorithm 5.

1: $r \leftarrow b - A \cdot x$
2: $p \leftarrow r$
3: $rr \leftarrow \langle r, r \rangle$
4: $bb \leftarrow \langle b, b \rangle$
5: $k \leftarrow 0$
6: **while** $\sqrt{rr} > \varepsilon \sqrt{bb}$ **do**
7:     $k \leftarrow k + 1$
8:     $q \leftarrow A \cdot p$
9:     $pq \leftarrow \langle p, q \rangle$
10:     $\alpha \leftarrow rr / pq$
11:     $x \leftarrow x + \alpha \cdot p$
12:     $r \leftarrow r - \alpha \cdot q$
13:     $rrold \leftarrow rr$
14:     $rr \leftarrow \langle r, r \rangle$
15:     $\beta \leftarrow rr / rrold$
16:     $p \leftarrow r + \beta \cdot p$
17: **end while**

The convergence of the iterative CG-solver depends on the condition number of A, i.e., $\kappa(A) = \lambda_{\max} / \lambda_{\min}$ where $\lambda_{\min}, \lambda_{\max}$ are the minimal and maximal eigenvalues of A, respectively. There holds the following upper bound for the convergence [9, Theorem 6.29]

$$\|x - x^{(k)}\|_A \leq 2 \left(\frac{\kappa(A)^{\frac{1}{2}} - 1}{\kappa(A)^{\frac{1}{2}} + 1} \right)^k \|x - x^{(0)}\|_A, \qquad (29)$$

using the $\|x\|_A^2 = \langle Ax, x \rangle$-norm. We observe that the CG-algorithm above requires the computation of two inner products, three vector operations and one Matrix–vector multiplication per iteration. Due to (29) we usually can obtain an accurate enough solution of the linear system with a number of iterations small compared with N.

5. Adaptivity, error estimators

If we know that the solution u of our variational problem in (4) is of higher regularity, then we can use Lemma 1 and one of the approximation results to prove convergence of the Galerkin method (7) and also determine the convergence rate. This result only shows asymptotic convergence and does not tell us any absolute errors, because the constants in the estimate are not known or have large values. For practical applications we need to control the global error, but we also want to have local error estimates. These error estimates have to be computable, i.e., they can not depend on the exact solution u, only on the Galerkin solution u_h, the right hand side functional, e.g., f, the domain, the mesh and any parameters in the PDE itself.

There are many different types of error estimators available in the literature. One of the simplest is the residual error estimator in Theorem 9. The error estimator alone only gives a computable upper bound for the error, but together with a mesh-refining strategy it can be used to create a sequence of adaptively refined meshes. For a convergence proof see, e.g., Ref. [10].

Let ω_h be a conforming mesh and let $\mathcal{E}_h$ the set of all edges in ω_h.

We will use the following neighbourhoods

$$\tilde{\omega}_T = \bigcup \{T' \in \omega_h \,;\, T \cap T' \neq \emptyset\}, \quad T \in \omega_h; \tag{30}$$

$$\omega_e = \bigcup \{T' \in \omega_h \,;\, e \in \partial T'\}, \quad e \in \mathcal{E}_h. \tag{31}$$

Theorem 9 (Residual error indicator). [11, *Proposition 4.2*] *Let ω_h be a quasi-uniform triangulation of Ω. Let u be the solution of the model problem in Example 1 and let u_h be the Galerkin solution. Then there exists a constant C, depending on Ω and ω_h, such that*

$$\|u - u_h\|_{H^1(\Omega)} \leq C\eta, \qquad \eta := \left(\sum_{T \in \omega_h} \eta_T^2 \right)^{1/2}, \tag{32}$$

with

$$\eta_T := \left(h_T^2 \|\Delta u_h + f\|_{L^2(T)}^2 + \frac{1}{2} \sum_{e \in \partial T} h_e \|[\frac{\partial u_h}{\partial n}]\|_{L^2(e)}^2 \right)^{1/2} \quad \forall T \in \omega_h. \tag{33}$$

Proof. We use a duality argument, i.e., we use

$$\|\nabla u - \nabla u_h\|_{L^2(\Omega)} = \sup_{w \in H_0^1, w \neq 0} \frac{(\nabla(u - u_h), \nabla w)_{L^2(\Omega)}}{\|\nabla w\|_{L^2(\Omega)}}. \tag{34}$$

Now, we apply integration by parts to $(\nabla(u - u_h), \nabla w)_{L^2(\Omega)}$ element wise, i.e.,

$$\langle l, w \rangle := (\nabla(u - u_h), \nabla w)_{L^2(\Omega)}$$

$$= (f, w)_{L^2(\Omega)} - \sum_{T \in \omega_h} (\nabla u_h, \nabla w)_{L^2(T)}$$

$$= (f, w)_{L^2(\Omega)} - \sum_{T \in \omega_h} \left\{ (-\Delta u_h, w)_{L^2(T)} + \sum_{e \in \partial T} (\nabla u_h \cdot n, w)_{L^2(e)} \right\}$$

$$= \sum_{T \in \omega_h} (\Delta u_h + f, w)_{L^2(T)} + \sum_{\substack{e \in \mathcal{E}_h \\ e \cap \partial \Omega = \emptyset}} \left(\left[\frac{\partial u_h}{\partial n} \right], w \right)_{L^2(e)}. \tag{35}$$

Due to the approximation theorem of Clément [12] there exists for every $w \in H_0^1(\Omega)$ an element $I_h w \in V_h$ such that

$$\|w - I_h w\|_{L^2(T)} \leq c\, h_T \|\nabla w\|_{L^2(\tilde{\omega}_T)} \quad \text{for } T \in \omega_h, \tag{36}$$

$$\|w - I_h w\|_{L^2(e)} \leq c\, h_e^{1/2} \|\nabla w\|_{L^2(\omega_e)} \quad \text{for } e \in \mathcal{E}_h. \tag{37}$$

The $\tilde{\omega}_T, T \in \omega_h$ define a finite covering of Ω. Therefore, we have with the Galerkin orthogonality

$$(\nabla(u - u_h), \nabla v_h)_{L^2(\Omega)} = 0 \quad \text{for } v_h \in V_h, \tag{38}$$

that there holds

$$\langle l, w \rangle = \langle l, w - I_h w \rangle$$

$$\leq \sum_{T \in \omega_h} \|\Delta u_h + f\|_{L^2(T)} \|w - I_h w\|_{L^2(T)}$$

$$+ \sum_{\substack{e \in \mathcal{E}_h \\ e \cap \partial \Omega = \emptyset}} \left\| \left[\frac{\partial u_h}{\partial n} \right] \right\|_{0,e} \|w - I_h w\|_{L^2(e)}$$

$$\leq c \sum_{T \in \omega_h} h_T \|\Delta u_h + f\|_{L^2(T)} \|\nabla w\|_{L^2(T)}$$

$$+ c \sum_{\substack{e \in \mathcal{E}_h \\ e \cap \partial \Omega = \emptyset}} h_e^{1/2} \left\| \left[\frac{\partial u_h}{\partial n} \right] \right\|_{L^2(e)} \|\nabla w\|_{L^2(\omega_e)}$$

$$\leq c \sum_{T \in \omega_h} \eta_T \|\nabla w\|_{L^2(T)} \leq c\eta \|\nabla w\|_{L^2(\Omega)}. \tag{39}$$

$\square$

Adaptive algorithm

For $k \in \mathbb{N}_0$ let ω_k be a regular triangulation of Ω. Let $u_k \in V_k \subset H_0^1(\Omega)$ denote the Galerkin solution. We define $N_k := |\omega_k|$. First, we give the algorithm for a shape-regular refinement scheme.

Algorithm 6. [13] Let ω_k be a conforming mesh. Let $\mathcal{E}_k$ be the set of all edges in ω_k. The edges of every triangle $\delta \in \omega_k$ are denoted by $e_{\delta,1}, e_{\delta,2}, e_{\delta,3}$ and the longest edge of $\delta \in \omega_k$ by e_δ. With $E_0 \subset \mathcal{E}_k$ we denote an initial set of edges, which have to be refined. Set $i = 0$.

(1) Set $i \leftarrow i + 1$.
(2) E_i is defined by

$$E_{i-1} \subseteq E_i.$$

For all $\delta \in \omega_k$ do

If $\{e_{\delta,1}, e_{\delta,2}, e_{\delta,3}\} \cap E_i \neq \emptyset$ then let $e_\delta \in E_i$.

end do

(3) if $E_i \neq E_{i-1}$ then goto 1.

Now, for all elements $\delta \in \omega_k$ which have at least one edge in E_i, the longest edge is also in E_i. The new mesh ω_{k+1} is defined as follows:

1: **for all** $\delta \in \omega_k$ **do**
2:     **if** $\{e_{\delta,1}, e_{\delta,2}, e_{\delta,3}\} \cap E_i = \{\emptyset\}$ **then**
3:         $\delta \in \omega_{k+1}$
4:     **else**
5:         connect the midpoint of the longest edge e_δ with the opposite vertex.
6:         **for all** $e \in \{e_{\delta,1}, e_{\delta,2}, e_{\delta,3}\} \cap E_i$ with $e \neq e_\delta$ **do**
7:             connect the midpoint of e_δ with the midpoint of e, cf. Figure 2.
8:         **end for**
9:         This creates 2, 3 or 4 new elements in ω_{k+1}.
10:     **end if**
11: **end for**

Due to [13] the newly constructed mesh ω_{k+1} is a conforming triangulation with the following characteristics:

(1) All edges in E_0 have been refined in ω_{k+1}.
(2) ω_{k+1} is nested in ω_k in such a way that each refined triangle is embedded in one triangle of ω_k in one of the ways shown in Figs. 2 and 3.

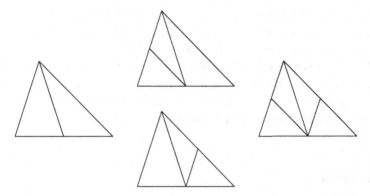

Fig. 2.   Refinement of $\delta \in \omega_k$. The new nodes are the midpoints of the edges of δ.

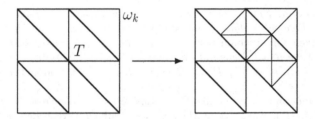

Fig. 3.   Local refinement of $T \in \omega_k$ and of the neighbour elements.

(3) The triangulation ω_{k+1} is non-degenerate; namely, the interior angles of all triangles of ω_{k+1} are guaranteed to be bounded away from 0.

(4) ω_{k+1} is smooth, in the sense that the transition between large and small triangles is not abrupt.

We estimate the global error, see (32), by

$$\eta_k = \left(\sum_{T \in \omega_k} \eta_{T,k}^2 \right)^{1/2}. \tag{40}$$

The adaptive refinement algorithm reads as follows:

Algorithm 7. Let the parameters $\theta \in [0,1]$, $\delta > 0$ and an initial triangulation ω_0 of Ω be given. With V_0 we denote the initial test and trial space, which is induced by the mesh ω_0. For $k = 0, 1, 2, \ldots$ perform

(1) Compute the solution $u_k \in V_k$ of the Galerkin scheme (7).

(2) Compute the error indicators $\eta_{i,k}$ according to Theorem 9. Compute the global error estimate η_k. Stop if $\eta_k < \delta$.

(3) The refinement set E_0 contains the edges of triangle $T \in \omega_k$ if for the local error indicator holds $\eta_{T,k} \geq \theta \max_{T' \in \omega_k} \eta_{T',k}$. Apply Algorithm 6 to (ω_k, E_0) to create a new and conforming mesh ω_{k+1}, where all edges in E_0 have been refined. This also defines the enlarged space $V_{k+1} \supset V_k$. Goto 1.

References

[1] D. Braess, *Finite Elements. Theory, Fast Solvers, and Applications in Solid Mechanics*, Cambridge University Press, Cambridge, 2001.

[2] S. Brenner and R. Scott, *The Mathematical Theory of Finite Element Methods*, Springer-Verlag, New York, 2008.

[3] C. Johnson, *Numerical Solution of Partial Differential Equations by the Finite Element Method*, Cambridge University Press, Cambridge, 1990.

[4] A. Aitken, On bernoulli's numerical solution of algebraic equations, *Proceedings of the Royal Society of Edinburgh.*, **46** (1926), 289–305.

[5] C. Schwab, *p- and hp-Finite Element Methods.*, Clarendon Press, Oxford, 1998.

[6] A. Kufner and A.-M. Sändig, *Some Applications of Weighted Sobolev Spaces.* Vol. 100, *Teubner-Texts in Mathematics*, Teubner-Verlag, Leipzig, 1987.

[7] R. A. Adams, *Sobolev Spaces.*, Academic Press, New York, 1975.

[8] B. Q. Guo and I. Babuška, The h–p version of the finite element method, part 1: The basic approximation results, *Comp. Mech.*, **1** (1986), 21–41.

[9] Y. Saad, *Iterative Methods for Sparse Linear Systems.* SIAM (2003). Available at http://www-users.cs.umn.edu/~saad/IterMethBook_2ndEd. pdf. Last accessed July 2015.

[10] W. Dörfler and R. Nochetto, Small data oscillation implies the saturation assumption, *Numer. Math.*, **91**(1) (2002), 1–12.

[11] R. Verfürth, A posteriori error estimation and adaptive mesh-refinement techniques, *J. Comput. Appl. Math.*, **50** (1994), 67–83.

[12] P. Clément, Approximation by finite element functions using local regularisation, *R.A.I.R.O. Anal. Numer.*, **9** (1975), 77–84.

[13] M.-C. Rivara, Mesh refinement processes based on the generalized bisection of simplices, *SIAM J. Numer. Anal.*, **21** (1984), 604–613.

Chapter 5

Introduction to Random Matrix Theory

Igor E. Smolyarenko

Department of Mathematics, Brunel University,
Kingston Lane, Uxbridge, UB8 3PH, UK
igor.smolyarenko@brunel.ac.uk

These notes provide a basic introduction to the main ideas and methods of Random Matrix Theory, aimed at a broad audience of scientists and engineers. Most of the technical details are restricted to the simplest case of unitary symmetry.

1. Introduction

1.1. *Historical sketch*

As implicit in the name, random matrix theory (RMT) is concerned with the study of matrices whose elements are random variables. The first occasion where an object that is now recognized as a random matrix made an appearance was the work of J. Wishart [1] on multivariate statistics, generalising the χ^2-distribution. At that time, though, it did not enter the mathematics mainstream, and the concept of a random matrix was reinvented (and for the first time so named) in the 1950's by nuclear physicists [2] concerned with the study of high-lying eigenstates of atomic nuclei. RMT was further developed and put on a reasonably firm mathematical footing by Mehta [3], Gaudin [4], and Dyson [5–7].

In the subsequent years, development of RMT has been largely driven by applications, mostly in physics at first, and later on in other natural and social sciences. In the early 80's, a generalisation of experimental and numerical observations led to the formulation of the

139

so-called Bohigas–Giannoni–Schmit (BGS) conjecture [8] (see also an earlier precursor in Ref. [9]) establishing a deep relation between random matrices and chaotic dynamical systems. Other instances where random matrices emerged as useful modelling tool span the range from quantum physics [10] to crystallography [11, 12], stock markets [13, 14], medical sciences [15], wireless communications [16] and complex networks [13]. In the early 1970's, H. Montgomery [17, 18] brought to light a connection between random matrices and number theory — non-trivial zeros of a broad class of *L*-functions asymptotically obey statistics of random matrix spectra.[a] Reflecting its origins and uses in applied sciences, RMT literature is exceptionally heterogeneous, with varying standards of mathematical rigour, and widely disparate 'languages'. In recent years, however, driven partly by the internal logic of the development of the probability theory, RMT underwent explosive growth, and became better incorporated into the main body of modern mathematics [13, 21–23].

The variety of settings in which RMT is applied leads to a rich variety of the types of random matrices, types of questions being asked of the theory, and, therefore, a broad range of mathematical tools being used (and sometimes specially developed!) to answer them. To name but a few, orthogonal polynomials [3], Riemann–Hilbert problems in complex analysis [24], analysis on graded symmetric spaces [25, 26], and free probability [27] all make appearances in random matrix research.

1.2.   *Universality: General remarks*

Whereas the work of Wishart [1] was a technical advance dealing with distributions of multicomponent samples taken from ordinary stochastic processes, the introduction of RMT by Wigner and Dyson was a more radical step. About a hundred years prior, statistical mechanics emerged as the result of a break with classical determinism, trading the concept of calculating many-particle deterministic classical trajectories for a statistical description of them. While the trajectories were thought of as governed by immutable Hamiltonian dynamics, averaging over initial conditions was sufficient to enable a statistical description (for reasons which became clear only much later with the development of the theory of chaotic dynamical

---

[a]This fact strengthens the case for the existence of the conjectural (see Refs. [19] and [20], though) Hilbert–Polya operator, the self-adjointness of which would imply the truth of the Riemann conjecture.

systems). The radical proposal of Wigner and Dyson was to treat the Hamiltonian itself (now in the quantum context) as a stochastic object. The essence of the conjecture was that, since the matrix elements of the Hamiltonian describing a large nucleus are determined in a very complicated way by interactions involving dozens or hundreds of nucleons, it is not unreasonable to model them as random variables.

The enormous success of this enterprise hinges on the *universality* property, central to RMT. Loosely, it can be formulated as follows: statistical properties of the spectra of *large* random matrices are independent of the specific details of the underlying distributions of stochastic matrix elements, and fall into a finite number of *universality classes* determined only by certain discrete symmetry constraints that may be imposed on the distribution. Thus, spectral statistics is different for random real symmetric matrices compared to random complex Hermitian matrices. 'Large' in this context means limiting behaviour as the dimension N of the matrix tends to infinity. The goal of this chapter is to provide an elementary introduction to the basic ideas and mathematical tools involved in the study of these universal properties. With one exception, consideration here is restricted to Hermitian matrices.

As the concept of universality plays a central role, it is worth putting it in the larger context of probability theory. Most often, a random matrix model is defined by postulating the joint distribution of all matrix elements. The object of study, typically, is the eigenvalue spectrum of the matrix as its dimension N becomes large (both the $N \to \infty$ limit and finite-N corrections may be of interest). The core reason why each model is not *sui generis* is similar to the mechanism underlying the classical Central Limit Theorem. A version of the latter states that, provided they possess two finite moments, the (appropriately scaled!) sum of a collection of random variables converges in distribution to the normal distribution. The salient for the present discussion features of this theorem are the following: (1) the subject of the theorem is a *single* random variable constructed out of a collection of $N \to \infty$ random variables, thus effecting a reduction by a factor of $O(N)$ of the number of variables, and (2) convergence to a unique distribution is ensured by rescaling the summed random variable by the square root of its (N-dependent) variance. In the case of large random matrices, the set of N eigenvalues of a matrix is obtained from $O(N^2)$ random variables characterising the elements of the matrix, effecting an $O(N)$ reduction in the number variables. Furthermore, rescalings exist that collapse various statistical measures of the eigenvalue spectra onto non-trivial universal functions. Considered from this point of view, RMT is a collection of similarly reductive laws governing the statistical properties of the spectral variables (or, more generally, of the eigenstructure) of large matrices whose entries are random variables.

2. Toy models

Although our main interest is in the behaviour of large random matrices, the key statistical property of eigenvalues induced by stochasticity of matrix elements can be introduced using 'toy models' of 2×2 matrices with stochastic entries. Consider an ensemble of real symmetric 2×2 matrices $H = \begin{pmatrix} a & b \\ b & c \end{pmatrix}$, where a, b and c are scalar real random variables. Previewing the fact that normal distributions of matrix elements will play an important role later, we assume the following joint probability density: $P_1(a, b, c) = \frac{1}{2\pi^{3/2}} e^{-\frac{1}{2} \operatorname{Tr} H^2}$. The eigenvalues of H are $\lambda_{1,2} = E \pm R$, where $E = (a+c)/2$, $R^2 = b^2 + \Omega^2/4$, and $\Omega = a - c$. Let us now consider the marginal joint distribution of the eigenvalues:

$$ \mathcal{P}_1(\lambda_1, \lambda_2) = \frac{1}{2} \int P_1(a, b, c) \delta(\lambda_1 - E - R) \delta(\lambda_2 - E + R) da\, db\, dc, \quad (1) $$

where $\delta(\cdot)$ is the Dirac delta-function, and the factor $1/2$ accounts for the ordering. Much of the technical challenge in RMT stems from the need to calculate multiple integrals, and hence to choose the appropriate parametrisation of the integration variables. One obvious way to do this here to change the variables of integration from (a, c) to (E, Ω), and then to polar coordinates (R, ϕ) in the plane $(\Omega/2, b)$. In terms of the new coordinates, $\frac{1}{2} \operatorname{Tr} H^2 = E^2 + \frac{1}{4}\Omega^2 + b^2 = E^2 + R^2$, and the measure is $da\, db\, dc = 2R dE\, dR\, d\phi$. Note that we have arrived at this choice of variables through prior knowledge of the explicit solution for the eigenvalues (λ_1, λ_2) in terms of the original matrix elements (a, b, c). Such information is not available in the case of large matrices (if only because of the need to solve high-order polynomial equations). However, examining the resulting measure, we notice that $2R = |\lambda_1 - \lambda_2|$, $2E = \lambda_1 + \lambda_2$, and H is diagonalized by the orthogonal rotation UHU^T, with $u = \begin{pmatrix} \cos\phi/2 & \sin\phi/2 \\ -\sin\phi/2 & \cos\phi/2 \end{pmatrix}$, $U^T U = \mathbb{I}$. Therefore, our change of variables neatly separates the spectral and angular degrees of freedom of H. Since our choice of P_1 depends only on the matrix invariant $\operatorname{Tr} H^2$, the integration over angular degrees of freedom becomes trivial, and the resulting marginal distribution of the eigenvalues is

$$P_1(\lambda_1, \lambda_2) = \frac{1}{4\sqrt{\pi}}|\lambda_1 - \lambda_2|e^{-\frac{1}{2}(\lambda_1^2 + \lambda_2^2)}. \tag{2}$$

The key feature of this result is the linear decay of the joint distribution as the two eigenvalues approach each other. This decay is the single most important fact about random matrix theory. Fundamentally, this is a geometric phenomenon: in order for the two eigenvalues to coincide, the discriminant $(2R)^2$ has to vanish, which is only possible when both b and $a - c$ are simultaneously equal to 0. In the three-dimensional space (a, b, c) this happens along a one-dimensional subspace spanned by E. As underscored by the change to the polar coordinates in the $\left(\frac{a-c}{2}, b\right)$ plane, this decay is analogous to the distribution of the distance from a random point uniformly distributed over a disk to the centre of the disk: the probability for it to lie at a distance $x \in (r, r + dr)$ from the centre is proportional to the area of the annulus of radius r and width dr, and thus proportional to r, vanishing as $r \to 0$.

Crucially, this linear decay is inherited not from any features of P_1, but from the Jacobian $2R = |\lambda_1 - \lambda_2|$. Moreover, as will be seen below, the Jacobians generated by the change of variables from matrix elements to spectral and angular degrees of freedom can be calculated for matrices of arbitrary dimension *without the intermediate step of explicitly evaluating the eigenvalues*. Whereas solving for the eigenvalues is equivalent to finding the full nonlinear diagonalising transformation U, the Jacobian follows from the simpler structure of the corresponding tangent space. In a sense, the remarkable power of RMT lies in its ability to extract statistical information about eigenvalues while bypassing explicit solutions of the characteristic equations in terms of the matrix elements.

Let us now repeat this exercise for a 'toy model' of 2×2 random complex Hermitian matrices, $H = \begin{pmatrix} a & b_0 + ib_1 \\ b_0 - ib_1 & c \end{pmatrix}$, where a, b_0, b_1, and c are scalar random variables. We again assume that the matrix elements are distributed normally and independently: $P_2(a, b_0, b_1, c) = \frac{2}{\pi^2}e^{-\text{Tr }H^2}$. The eigenvalues of H are now $\lambda_{1,2} = E \pm R$, where $R^2 = b_0^2 + b_1^2 + \Omega^2/4$, and, as before, $E = (a+c)/2$, and $\Omega = a-c$. The change of variables $(a, c) \to (E, \Omega)$, and then to spherical co-ordinates (R, θ, ϕ) in the $(\Omega/2, b_0, b_1)$ space gives $\text{Tr }H^2 = 2(E^2 + R^2)$, and

$$P_2(\lambda_1, \lambda_2) = \frac{1}{2} \int P_2(a, b_0, b_1, c)\delta(\lambda_1 - E - R)\delta(\lambda_2 - E + R)da\, db_0\, db_1\, dc$$

$$= 2 \int dE\, R^2 dR \sin\theta\, d\theta\, d\phi\, e^{-2E^2 - 2R^2}\delta(\lambda_1 - E - R)\delta(\lambda_2 - E + R)$$

$$= \frac{1}{\pi}(\lambda_1 - \lambda_2)^2 e^{-\lambda_1^2 - \lambda_2^2}. \tag{3}$$

The decay of the joint distribution function as $\lambda_2 \to \lambda_1$ is now quadratic, due to the fact that *three* random variables $(a - c)$, b_0, and b_1 have to vanish simultaneously in order for the two eigenvalues to be degenerate. The geometric analogy is to the distribution of distances from a point uniformly seeded in a three-dimensional ball to its centre. The off-diagonal elements in these two 'toy' examples were real and complex numbers, respectively. The only possible generalisation that preserves the associativity of the matrix product is that of quaternion matrices:

Exercise 2.1. *Consider a 2×2 matrix of real quaternions which can be realised as a 4×4 matrix with the following block structure:*

$$H = \begin{pmatrix} ae_0 & b_0 e_0 + \mathbf{b} \cdot \mathbf{e} \\ b_0 e_0 - \mathbf{b} \cdot \mathbf{e} & ce_0 \end{pmatrix}, \tag{4}$$

where the quaternion basis is given by

$$e_0 = \begin{pmatrix} 1 & 0 \\ 0 & 1 \end{pmatrix}; \quad e_1 = \begin{pmatrix} i & 0 \\ 0 & -i \end{pmatrix}; \quad e_2 = \begin{pmatrix} 0 & 1 \\ -1 & 0 \end{pmatrix}; \quad e_3 = \begin{pmatrix} 0 & i \\ i & 0 \end{pmatrix} \tag{5}$$

and a, c, and $b_{0...3}$ are six real parameters. It is easy to see that the matrix is Hermitian by construction. Show that the eigenvalues are doubly degenerate, and that the joint distribution of the two distinct eigenvalues exhibits quartic decay. **Hints:** *(1) Show that e_is satisfy the rules of quaternion algebra. (2) Show that $[H, T] = 0$, where $T = UK$, $U = e_2 \otimes \mathbb{I}$, $\mathbb{I}$ is the unit matrix in the 2×2 block subspace, and K effects complex conjugation. Hence show that for any eigenvector $\mathcal{V}$ of H, written in the block notation as $\mathcal{V} = \begin{pmatrix} v_1 \\ v_2 \end{pmatrix}$, the vector $T\mathcal{V}$ is also an eigenvector of H. (In quantum-mechanical applications, T is the time-reversal operator.) (3) Show that $\mathcal{V}$ and $T\mathcal{V}$ are orthogonal, and hence show that the eigenvalues of H are doubly degenerate (special case of the Kramers' theorem). (4) Use the block structure of $\mathcal{V}$ to derive and solve a quadratic equation on the two distinct eigenvalues $\lambda_{1,2}$ of H. (5) Perform the change of variables from $\{a, c, b_0, \mathbf{b}\}$ to λ_1, λ_2 and angular variables, and hence show that the joint distribution of λ_1 and λ_2 is proportional to $(\lambda_1 - \lambda_2)^4$.*

2.1.   *Universality: Eigenvalue repulsion*

In the three 'toy' examples, the joint distribution of the two eigenvalues decays asymptotically according to $|\lambda_1 - \lambda_2|^\beta$ as $\lambda_1 \to \lambda_2$, where $\beta = 1, 2, 4$, for real symmetric, complex Hermitian, and real quaternion matrices. The vanishing of the probability for two eigenvalues to be in close proximity is often called 'eigenvalue repulsion'. The three integer values of β correspond to the number of components in the three possible associative number systems, and follow from the fact that in the space of all parameters characterising the matrix, the condition for two eigenvalues to coincide is fulfilled in subspaces of co-dimension $C = 2, 3$, and 5 in the cases of $\beta = 1, 2, 4$, respectively.

The specific form of the distributions P_β is (almost) irrelevant. For example, in the $\beta = 1$ case, the probability that the two eigenvalues are within a small distance $\epsilon > 0$ from each other is the probability that the square root of the discriminant $2R$ of the corresponding quadratic equation falls within that range. This is given by $\int_{2R \le \epsilon} P_1(a, b, c) da\ db\ dc$, assuming now that P_1 is arbitrary. If P_1 is a smooth function near $b = 0$ and $a = c$, the integral is equal to $\frac{1}{2} L(0,0) \epsilon^2 + O(\epsilon^3)$, where $L(\Omega, b) = \pi \int dE P_1(E + \Omega/2, b, E - \Omega/2)$, with the corresponding probability density equal to $L(0,0) \epsilon$ to the leading order. In order to 'destroy' this behaviour one would need either $L(0,0)$ to vanish, or $L(\Omega, b)$ to be singular near the origin. While not impossible, in most sensible models the distributions of matrix elements have no 'reason' to exhibit special behaviour just where the eigenvalue distance vanishes. The same reasoning extends to the other two values of β where the leading terms of the probability density are easily seen to be $O(\epsilon^\beta)$. On the other hand, the coefficient of proportionality $L(0,0)$ seems to depend substantively on the specifics of P_1 in the 2×2 case. It will be seen later that the corresponding behaviour in the $N \to \infty$ case is controlled by a *single parameter* equal to the analogue of $\mathbb{E}[|\lambda_1 - \lambda_2|]$ (*mean eigenvalue spacing* conventionally denoted as Δ). This is the rescaling parameter that is usually necessary to elicit universal functional forms, as mentioned earlier.

The previous comments naturally lead to inquire about the relevance of these 'toy' 2×2 models to the behaviour of general $N \times N$ matrices. Consider the analogue of the L integral above, determining the probability that two neighbouring eigenvalues happen to lie within a small distance ϵ from each other. The integral now involves all random variables characterising the $N \times N$ matrix. Let us consider a change of variables effecting a partial diagonalisation such that the matrix acquires a block-diagonal structure, with a 2×2 block corresponding to the two close eigenvalues, and the remaining $(N-2) \times (N-2)$ block. The variables in the latter block do not affect the two eigenvalues of interest. *If* it can be reasonably assumed that the Jacobian does not contain a sharp dependence on the difference between these two eigenvalues, *and* that the integration over the independent variables in the diagonalising matrices does not introduce one either, the remaining integration has the structure of the L integral in the leading-order, thus enforcing the ϵ^β behaviour. The restriction of the values of the eigenvalue repulsion exponent to the three values $\beta = 1$, 2, 4 (Dyson's '3-fold way' [7]) has deep algebraic reasons, and is one of the central features of RMT universality.

3. Random matrix ensembles

3.1. *Why Gaussian?*

The starting point for constructing a random matrix model is choosing the *ensemble* — the joint distribution of all elements of the matrix. The two most commonly made assumptions are (1) that the distribution should be invariant with respect to the change of basis, and (2) that each matrix element is independently distributed. Both of these assumptions can be viewed as expressions of the lack of information about the system. Either one, taken independently, defines an important class of ensembles.

We will now show that, if both assumptions are enforced, the form of the distribution of the matrix elements is severely restricted [3, 28]. Consider, for definiteness, random complex $N \times N$ Hermitian matrices. The measure on the space of such matrices is $dH = \prod_{i=1}^{N} dH_{ii} \prod_{i<j}^{N} dH_{ij}^{(0)} dH_{ij}^{(1)}$, where $H_{ij}^{(0)} = \Re H_{ij}$ and $H_{ij}^{(1)} = \Im H_{ij}$. A distribution $P(H)$ is invariant if $P(H)dH = P(\tilde{H})d\tilde{H}$, where $\tilde{H} = UHU^{-1}$ for any unitary matrix U. It is straightforward to show that $dH = d\tilde{H}$. One possible route is to use the fact that, since a Hermitian $N \times N$ matrix is described by N^2 real parameters, $\mathcal{S}(H) \approx \mathbb{R}^{N^2}$, where $\mathcal{S}(H)$ is the space of all Hermitian matrices H. Thus, a unitary transformation of H is an orthonormal transformation with respect to a certain inner product on $\mathbb{R}^{N^2}$ (see Chapter 5 of Ref. [24]). The requirement of invariance therefore implies $P(H) = P(\tilde{H})$. The assumption of independence means that $P(H)$ factorises:

$$P(H) = \prod_{i=1}^{N} p_{ii}(H_{ii}) \prod_{i<j} p_{ij}^{(0)} \left(H_{ij}^{(0)} \right) p_{ij}^{(1)} \left(H_{ij}^{(1)} \right). \tag{6}$$

In order to ensure invariance of P under unitary rotations, it is enough to ensure invariance under infinitesimal rotations $U = e^{i\epsilon A}$, where $A^\dagger = A$, and $\epsilon \to 0$. Invariance of P implies $\frac{d}{d\epsilon}P(\tilde{H}) = 0$. Evaluating the derivative of P and using $\frac{d\tilde{H}}{d\epsilon} = i[A, \tilde{H}]$, we find

$$\sum_{i=1}^{N} \frac{d\ln p_{ii}}{dH_{ii}} \left[A, \tilde{H}\right]_{ii} + \sum_{\alpha=0,1} \sum_{i<j} \frac{d\ln p_{ij}^{(\alpha)}}{dH_{ij}^{(\alpha)}} \left[A, \tilde{H}^{(\alpha)}\right]_{ij} = 0. \tag{7}$$

The structure of this equation is such that each term in the sum is a product of a (possibly nonlinear) function depending only on a single component of H, and an expression linear in the components of H. Therefore, quadratic and higher-order contributions in each term with the structure $(d\ln p_\mu/dH_\mu)H_{\mu'}$ have to be cancelled (up to a constant) by a dual term $(d\ln f_{\mu'}/dH_{\mu'})H_\mu$, where μ is a short-hand for an independent combination of i, j, and (α). Such a cancellation is only possible if the logarithmic derivatives of p are no more than linear in their unique variables. Therefore, $\ln P(H)$ is no more than quadratic in the invariants, and the possible term $(\text{Tr}H)^2$ is forbidden by the requirement of independence. Hence, $\ln P(H)$ is restricted to the form $\left\{-\alpha_1 \text{Tr}\,H^2 + \alpha_2 \text{Tr}H + \alpha_3\right\}$, with arbitrary constants $\alpha_{1...3}$. This argument essentially eschews any reliance on the structure of $[A, H]$ apart from its linearity. A constant shift of H does not affect any statistical properties of the eigenvalues, so one could set $\alpha_2 = 0$. The parameter α_1 controls only the overall scale of the spectrum, and can be set arbitrarily, e.g., $\alpha_1 = 1$, thus defining the so-called *Gaussian Unitary Ensemble* (GUE):

$$P(H) = \mathcal{A}_2(N) \exp\{-\text{Tr}\,H^2\}, \tag{8}$$

where $\mathcal{A}_2(N) \equiv e^{\alpha_3}$ is the normalising constant.

Exercise 3.1. *Prove the Gaussian form of the distribution under the same two assumptions for real symmetric matrices (Gaussian Orthogonal Ensemble, GOE).*

The same statement can be proved for real quaternion matrices (Gaussian Symplectic Ensemble, GSE). A conventional choice is $\alpha_1 = \beta/2$, thus for all three ensembles $P(H) \propto \exp\left\{-(\beta/2)\text{Tr}\,H^2\right\}$.

3.2. *Dyson diffusion*

The special role played by the Gaussian distribution can be further elucidated from a somewhat different point of view, first developed by

Dyson [6], who considered a *diffusion process in the space of matrices*. Recall that a scalar Wiener process $W(t)$ is defined [29] as a continuous time Markov process with continuous sample paths, and independent increments $W(t+s) - W(t)$ distributed according to $N(0, \sigma^2 s)$, with some constant σ^2. An *Ornstein–Uhlenbeck* process $B(t)$ can be defined through $dB(t) = -\gamma B(t)dt + dW(t)$, with some positive constant γ. The role of the first term here is to provide a 'restoring force' which counteracts the diffusive spread of the Wiener process. The forward equation for the time-dependent density function $p(b, t)$ is

$$\frac{\partial p(b,t)}{\partial t} = \gamma \frac{\partial(bp(b,t))}{\partial b} + \frac{\sigma^2}{2} \frac{\partial^2 p(b,t)}{\partial b^2}. \tag{9}$$

The forward equation is solved by

$$p(b,t) = \sqrt{\frac{\gamma}{\pi \sigma^2 (1 - q^2(t))}} \exp\left\{ -\frac{\gamma \left[b - q(t)b_0\right]^2}{\sigma^2 [1 - q^2(t)]} \right\}, \tag{10}$$

where the initial value of the process is $B(0) = b_0$, and $q(t) = e^{-\gamma t}$. The solution converges to a b_0-independent stationary Gaussian distribution with variance σ^2/γ in the $t \to \infty$ limit.

Let us now consider [6] a matrix generalisation in which each matrix element H_μ executes an independent Ornstein–Uhlenbeck process with $\gamma_\mu = \gamma$, and $\sigma_\mu^2 = (1 + \delta_{ij})\sigma^2$. Then the time-dependent joint distribution $P(H, t)$ of all H_μ satisfies the forward equation

$$\frac{\partial P(H,t)}{\partial t} = \sum_\mu \left\{ \gamma \frac{\partial(H_\mu P(H,t))}{\partial H_\mu} + (1 + \delta_{ij}) \frac{\sigma^2}{2} \frac{\partial^2 P(H,t)}{\partial H_\mu^2} \right\}. \tag{11}$$

If $H = H_0$ at $t = 0$, the solution of this equation is a straightforward generalisation of Equation (10):

$$P(H,t) \propto \exp\left\{ -\frac{\gamma \mathrm{Tr}\left[H - q(t)H_0\right]^2}{\sigma^2 [1 - q^2(t)]} \right\}, \tag{12}$$

which converges to a Gaussian distribution in the $t \to \infty$ limit. There is some 'fine tuning' involved in singling out the diagonal terms in σ_μ^2 in order to ensure the invariance of the final result in Equation (12), however this is consistent with the observation that diagonal and off-diagonal components of H enter $\mathrm{Tr}H^2$ unequally.

3.3. *Types of random matrix ensembles*

The three types of associative number systems that exhaust the possibilities for matrix elements of Hermitian ($H^\dagger = H$) matrices correspond to real symmetric ($H^T = H$, $\beta = 1$), complex Hermitian ($\beta = 2$), and quaternion self-dual ($H^R \equiv e_2 H^T e_2^{-1}$, $\beta = 4$) matrices, generalising our 'toy' models [3]. (In quantum–mechanical applications the $\beta = 4$ case arises from the physics of time-reversal invariant half-integer spin systems, while the $\beta = 2$ case is distinguished by the absence of time-reversal invariance.)

Either of the two assumptions stated at the beginning of this section can be relaxed. If the independence requirement is dropped, one obtains *invariant* ensembles, represented (but not exhaustively) by distributions of the type $P(H) \propto e^{-\mathrm{Tr}V(H)}$. The function $V(x)$ is usually chosen to be a polynomial with an even senior term, but more general cases can also be considered. The three values of β correspond to distributions that are invariant under $H \mapsto U^{-1}HU$, where U is drawn from the orthogonal group for $\beta = 1$, unitary group for $\beta = 2$, and symplectic group for $\beta = 4$.

Dropping instead the invariance requirement defines *Wigner ensembles*: diagonal elements H_{ii} are independent identically distributed (iid), and the same is true for all components of the off-diagonal elements H_{ij}, with possibly a different distribution. The Gaussian ensembles are simultaneously also Wigner ensembles, with $\mathbb{E}[H_{ii}^2] = 1/\beta$, and also $\mathbb{E}[H_{ij}^2] = 1/2$ for GOE, and $\mathbb{E}[(\mathrm{Re}\,H_{ij})^2] = \mathbb{E}[(\mathrm{Im}\,H_{ij})^2] = 1/4$ for GUE.

If neither invariance nor independence is imposed, a variety of ensembles emerges. Sparse random matrices arise in physics: e.g., Anderson model of disorder in solid state physics $H = H_0 + D$, where H_0 is a (weighted) adjacency matrix of a large lattice graph, and D is a diagonal random matrix with iid elements [10]. Adjacency matrices in more general models of random graphs is an active area of current research. Non-Hermitian random matrices arise in various contexts, e.g., quantum chaotic scattering. Reference [13] provides an up-to-date survey. The consideration below (with minor exceptions) is limited to GUE which is technically among the simplest ensembles to study, allowing to introduce some core mathematical tools of RMT in an unencumbered setting.

3.4. *Eigenvalue repulsion: Dynamics*

Let us return to 'toy' 2×2 examples, and consider *non-Hermitian* matrices with the structure $A = \begin{pmatrix} a + E & b \\ -b & -a + E \end{pmatrix}$, drawn from the probability distribution $P_{nH}(a, b, E) = \frac{1}{\pi^{3/2}} e^{-\frac{1}{2} \operatorname{Tr} A^\dagger A}$, where the normalisation coefficient follows from $\operatorname{Tr} A^\dagger A = 2(a^2 + b^2 + E^2)$. The two eigenvalues of A are either real or complex conjugate depending on the sign of $a^2 - b^2$: $\lambda_{1,2} = E \pm \sqrt{a^2 - b^2}$.

Exercise 3.2. *Show that the join probability distribution of the two eigenvalues conditioned on their being real is*

$$\frac{1}{2\pi^{3/2}} |\lambda_1 - \lambda_2| e^{-(\lambda_1 + \lambda_2)^2/4} K_0 \left(\left[\frac{\lambda_1 - \lambda_2}{2} \right]^2 \right), \tag{13}$$

where K_0 denotes modified Bessel function of the zeroth order.
  Solution:
  The distribution of the eigenvalues conditioned on being real is

$$\mathcal{P}_{nH}(\lambda_1, \lambda_2) = \frac{1}{\pi^{3/2}} \int_{a^2 \geq b^2} da\, db\, dE\, e^{-a^2 - b^2 - E^2} \delta(\lambda_1 - E - R) \delta(\lambda_2 - E + R), \tag{14}$$

where $R = \sqrt{a^2 - b^2}$. Performing the trivial integration over E, and changing variables to $a = r \cosh t$, $b = r \sinh t$, we find

$$\mathcal{P}_{nH}(\lambda_1, \lambda_2) = \frac{4}{\pi^{3/2}} \int_0^\infty r\, dr \int_0^\infty dt\, e^{-2r^2 \cosh^2 t + 2\lambda_1 r - \lambda_1^2} \delta(\lambda_2 - \lambda_1 + 2r). \tag{15}$$

The factor r comes from the Jacobian $\left| \dfrac{\partial(a,b)}{\partial(r,t)} \right|$, and inasmuch as t could be considered to be the hyperbolic analogue of the angular variable ϕ in the parametrisation of the Hermitian case, it is the direct counterpart of the Jacobian factor in Hermitian models responsible for the vanishing of the probability of eigenvalue coincidence. We will see shortly, however, that in this non-Hermitian model the same linear decay of the Jacobian co-exists with the absence of dynamical eigenvalue repulsion! After integrating out the remaining delta-function, the integral reduces to

$$\tilde{\mathcal{P}}_{nH}(\lambda_1, \lambda_2) = \frac{1}{\pi^{3/2}} |\lambda_1 - \lambda_2| e^{-(\lambda_1 + \lambda_2)^2/4} \int_0^\infty dt\, e^{-\frac{(\lambda_1 - \lambda_2)^2}{4} \cosh 2t}. \tag{16}$$

The result quoted above follows from the definition of K_0.

Recalling that $K_0(z) \sim -\ln z$ as $z \to 0$, we see that decay of the joint eigenvalue density near $\lambda_1 = \lambda_2$ is logarithmically weaker than linear. Formally this logarithmic correction arises as a result of coupling between the 'angular variable' t and the eigenvalues via the Gaussian weight as seen in the integral (16) — in contrast to the factorisation in the invariant Hermitian case.

A crucial difference with the Hermitian examples considered before is that the co-dimension of the subspace where $\lambda_1 = \lambda_2$ is equal to 1. The condition for coincidence is $a^2 = b^2$, hence it is a two-dimensional surface in the three-dimensional space (a, b, E), as opposed to a 1-d line $a = b = 0$ in the real symmetric case. Consider, however, the (weighted by the distribution) volume of the ϵ-vicinity (defined *via* $|\lambda_1 - \lambda_2| < 2\epsilon$) of the intersection point in the space of parameters (a, b, E). In the Hermitian real symmetric case this volume is proportional to the volume of a ball of radius ϵ, i.e., $\pi \epsilon^2$ (assuming the distribution is smooth on the scale of ϵ near $a = b = 0$). In the non-Hermitian case, however, this volume is given by an integral over a union of wedges, narrowing towards infinity along $a = \pm b$ lines, which is proportional to $\epsilon^2 |\ln \epsilon|$ (the logarithmic correction stems from lack of invariance).

Armed with the concept of matrix diffusion, we can take another look at eigenvalue repulsion. Let individual independent matrix elements evolve as independent processes $da(t) = -a(t)dt + dW_a(t), db(t) = -b(t)dt + dW_b(t)$. In the real symmetric case, the distance between the eigenvalues is $|\lambda_1 - \lambda_2| = 2r$, where $r = \sqrt{a^2 + b^2}$. Using Ito's lemma [29] we obtain $dr^2(t) = 2\left[1 - r^2(t)\right] dt + 2a(t)dW_a(t) + 2b(t)dW_b(t)$. Since the processes $W_a(t)$ and $W_b(t)$ are assumed independent, this is equivalent to $dr^2(t) = 2\left[1 - r^2(t)\right] dt + 2r(t)dW(t)$, known as the two-dimensional squared Bessel process. Crucially, in the vicinity of $r = 0$, the dynamics is dominated by the $2dt$ term, i.e., by repulsion from zero.

In contrast, the difference between real eigenvalues in the non-Hermitian case is given by $r = \sqrt{a^2 - b^2}$, and application of Ito's lemma gives $dr^2(t) = -2r^2(t)dt + 2a(t)dW_a(t) - 2b(t)dW_b(t)$. There is weak *attraction* to zero due to the $r^2 dt$ term, but it is overwhelmed by the diffusion terms near $r^2 = 0$, resulting in overall 'uneventful' crossings of $r^2 = 0$. Note, however, that the crossing is more 'dramatic' if expressed in terms of r: $dr(t) = -dt \left(r + \frac{a^2 + b^2}{r^3}\right) + \frac{a}{r}dW_a - \frac{b}{r}dW_b$. The deterministic term is now singular near zero — the eigenvalues pass through the intersection point infinitely 'fast', ensuring that the dynamical 'time' spent in the ϵ-vicinity of the intersection is the same (up to logarithmic correction) in both real symmetric and real non-Hermitian cases. This behaviour generalises to large matrices (see, for example Ref. [30] for a recent study of the dynamics of real eigenvalues of non-Hermitian matrices), although it should be emphasised

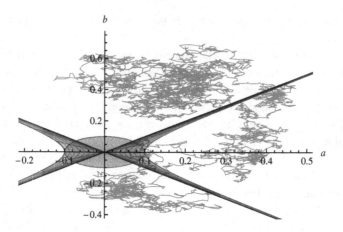

Fig. 1. A random path in the (a, b) plane, together with ϵ-vicinities of the eigen-value coincidence manifolds in Hermitian and non-Hermitian cases.

that our 2×2 non-Hermitian 'toy' is a non-invariant distribution, and the logarithmic correction to the repulsion strength does not extend to all eigenvalue pairs. Figure 1 shows a typical Brownian curve in the (a, b) plane together with the areas in which $|\lambda_1 - \lambda_2|$ is small in both Hermitian (real symmetric) and non-Hermitian cases.

4.   Spectral statistics in GUE

4.1.   *Distribution of eigenvalues*

The first step towards achieving a statistical description of the spectral variables of GUE is to derive the joint distribution of all eigenvalues. Essen-tially, one needs to change variables from the matrix elements of H to the eigenvalues and eigenfunctions, integrating over the latter. The key to the procedure is the calculation of the Jacobian. Suppose H is diagonalized by a unitary matrix U: $H = U \Lambda U^{-1}$, where $\Lambda = \text{diag}(\lambda_1, \lambda_2, \ldots, \lambda_N)$. It is worth noting here that the eigenvalues of a GUE matrix are generi-cally non-degenerate — the measure of the set of matrices with degenerate eigenvalues is 0 (see Chapter 5 of Ref. [24] for further details).

Consider the invariant distance ρ in the space of H: $d\rho^2 = \text{Tr} \, dH \, dH^\dagger$. Now,

$$dH = dU \Lambda U^{-1} + U \Lambda dU^{-1} + U d\Lambda U^{-1} = U \left([dA, \Lambda] + d\Lambda \right) U^{-1}, \quad (17)$$

where $dA = U^{-1}dU = -dU^{-1}U$. Therefore

$$d\rho^2 = 2\text{Tr}[dA\Lambda dA\Lambda] - 2\text{Tr}[dA^2\Lambda^2] + \text{Tr}[d\Lambda^2]$$
$$= -2\sum_{i<j} dA_{ij}dA_{ji}(\lambda_i - \lambda_j)^2 + \text{Tr}[d\Lambda^2]. \tag{18}$$

Counting the number of degrees of freedom (dA is complex!), we find that the Jacobian is given by the square of the *Vandermonde determinant*:

$$J = |\Delta_N(\{\lambda\})|^2 \equiv \prod_{1 \leq i < j \leq N} (\lambda_j - \lambda_i)^2. \tag{19}$$

Note that $\Delta_N(\{\lambda\}) = (-1)^{N(N-1)/2} \det_{1 \leq i < j \leq N} \lambda_j^{i-1}$. The joint distribution of the eigenvalues is obtained by integrating over the degrees of freedom of U. Since the Jacobian does not involve any dependence on U, the explicit form of the measure $dM(U)$ is not crucial, as long as one is not concerned with the calculation of the normalisation constant:

$$P(\{\lambda\}) = \mathcal{B}_2(N)|\Delta(\{\lambda\})|^2 e^{-\sum_{i=1}^N \lambda_i^2}. \tag{20}$$

The eigenvalue repulsion is now seen to be a generic feature involving all eigenvalues: the joint distribution function vanishes quadratically whenever any two eigenvalues coincide. This also implies that the eigenvalues are highly correlated. The existence of Equation (20), which greatly simplifies subsequent analysis of eigenvalue correlations is owed to the invariance of the distribution of matrix elements. No such expressions can be obtained for general Wigner ensembles which makes their analysis much more challenging. An analogous argument gives $P(\{\lambda\}) \propto |\Delta(\{\lambda\})|^\beta$ for the other two ensembles.

4.1.1. *Correlation functions*

Equation (20) effects a reduction in the number of independent variables from N^2 to N. Analysis of the correlations between the eigenvalues is enabled by further integrating out $O(N)$ eigenvalues leaving behind universal expressions containing $O(1)$ variables. There are two main classes of such objects: *correlation functions* defined below, which are functions of eigenvalues only, and expressions engaging the eigenvalue number — the simplest of those is the *spacing distribution*, defined as the distribution of distances between nearest eigenvalues. Let us concentrate first on correlation functions. The m-point correlation function R_m is defined as [3]

$$R_m(\lambda_1, \ldots, \lambda_m) = \frac{N!}{(N-m)!} \int P(\{\lambda\}) \prod_{j=m+1}^{N} d\lambda_j. \tag{21}$$

The simplest of these is the eigenvalue density:

$$R_1(\lambda) = N \int_{-\infty}^{\infty} P\left(\lambda, \{\lambda_j\}_{j=2}^{N}\right) \prod_{i=2}^{N} d\lambda_i. \tag{22}$$

The significance of R_1 is that for any set $B \in \mathbb{R}$, $\int_B R_1(x)dx = \mathbb{E}[\mathcal{N}_B]$, where $\mathcal{N}_B$ is the (random) number of eigenvalues in B.

4.2. *Wigner semicircle*

The multiple integral that needs to be evaluated in order to compute R_1 is an integral over all possible configurations of the N eigenvalues λ_j. The approach we take to calculating it is, in effect, a version of multidimensional saddle-point approximation: the integral is dominated, in the $N \to \infty$ limit, by the most probable configurations [3, 24]. In order to formalize this idea, let us re-scale the variables: $\lambda_j = \sqrt{N}\epsilon_j$, and introduce the density of scaled eigenvalues $\rho(x) = N \int \tilde{P}(\{\epsilon\})\delta(x-\epsilon_1) \prod_{j=1}^{N} d\epsilon_j$, where $\tilde{P}(\{\epsilon\}) \propto e^{-\mathcal{F}}$, with $\mathcal{F} = -\sum_{i\neq j}^{N} \ln|\epsilon_i - \epsilon_j| + N\sum_{i=1}^{N} \epsilon_i^2$. In terms of the new variables, it is easily seen that a change $O(1)$ in the value of a single ϵ_i involves a change $O(N)$ in the exponent, thus giving credence to the idea that the dominant contribution should come from the configurations which minimize $\mathcal{F}$.

Let us now introduce the normalised *counting measure* $\mu(x) = \sum_j \theta(x - \epsilon_j)/N$, in terms of which

$$\mathcal{F} = N^2 \left[-\int d\mu(x)d\mu(y)\ln|x-y| + \int x^2 d\mu(x) \right]. \tag{23}$$

The two crucial (and highly non-trivial) facts which we will state here without proof (see Ref. [24]) are that (i) the *equilibrium measure* $d\mu_0(x)$ exists and is given by the solution of the variational problem minimising $\mathcal{F}$, and (ii) $(1/N)\rho(x)dx$ converges to this equilibrium measure. The calculation of $\rho(x)$ thus reduces to the solution of a minimisation problem.

Assuming $d\mu_0(x) = \sigma(x)dx$ for some function $\sigma(x)$ with compact support, we find $-2\int dy\sigma(y)\ln|x-y| + x^2 = L$, where L is the Lagrange multiplier introduced in order to enforce the normalisation of the counting measure $\int d\mu_0(x) = 1$. It is convenient to differentiate this equation, so that

$$\mathcal{P} \int \frac{\sigma(y)dy}{x - y} = x, \tag{24}$$

where the symbol $\mathcal{P} \int$ denotes the principal value integral.

Equation (24) is of a type known as a *singular* integral equation. The standard method for analysing equations of this type is to reduce them to a *scalar* Riemann–Hilbert problem in complex analysis [24, 31] which can be solved in closed form. Let us introduce

$$F(z) = \frac{1}{i\pi} \int \frac{\sigma(y)}{y - z} dy, \tag{25}$$

which is analytic on $\mathbb{C}\backslash\Sigma$, where Σ is the support of $\sigma(x)$ on $\mathbb{R}$. We will assume the latter to be a symmetric interval $(-a, a)$. The value of a and the self-consistency of the assumption will be determined below. Denoting as $F_\pm(x)$ the limits of $F(z)$ above (below) Σ, we can rewrite the equation as a condition on $F_\pm(x)$ on Σ:

$$F_+(x) + F_-(x) = \frac{2ix}{\pi}. \tag{26}$$

A standard scalar Riemann–Hilbert problem on a contour Σ is defined as the problem of finding a function $f(z)$ such that it is (i) analytic on the complement of Σ, (ii) satisfies a 'jump condition' $f_+(z) = f_-(z)g(z)$ for some given function $g(z)$ defined on Σ, and (iii) $f(z) \to 1$ as $z \to \infty$. Taking the logarithm of the 'jump condition', it can be rewritten as

$$\ln f_+(z) - \ln f_-(z) = \ln g(z), \tag{27}$$

which is almost the form of Equation (26) except for the different sign in the l.h.s. Equation (27) is solved by the Plemelj formula

$$\ln f(z) = \int_\Sigma \frac{\ln g(\zeta)}{\zeta - z} \frac{d\zeta}{2\pi i}. \tag{28}$$

In order to apply Equation (28) in our case, we need to convert the sum in Equation (26) into a difference. This can be achieved by introducing an auxiliary function $R(z) = \sqrt{z^2 - a^2}$. It is evident that $R_+(x) = -R_-(x)$, and thus $G(z) = F(z)/R(z)$ satisfies a standard jump condition on Σ:

$$G_+(x) - G_-(x) = \frac{2ix}{\pi R_+(x)}. \tag{29}$$

The Plemelj formula can now be applied directly, leading to $G(z) = \frac{1}{\pi^2} \int_{-a}^{a} \frac{ydy}{(y-z)R_+(y)}$ and thus

$$F(z) = \frac{R(z)}{\pi^2} \int_{-a}^{a} \frac{ydy}{(y-z)R_+(y)}. \tag{30}$$

Compactness of the support of σ is responsible for the $F(z) \to 0$ as $z \to \infty$ condition in the formulation of the Riemann–Hilbert problem. Expanding $1/(y-z) = (-1/z)(1 + y/z + \cdots)$, we see that this condition is satisfied if

$$\int_{-a}^{a} \frac{xdx}{R_+(x)} = 0, \tag{31}$$

which, in turn, is guaranteed by our initial guess of a symmetric interval $(-a, a)$. Furthermore, recalling that the equilibrium measure must be normalised, we need to ensure $\int_{-a}^{a} dx\sigma(x) = 1$. Noting that as $z \to \infty$, $F(z) = -\frac{1}{i\pi z} \int_{-a}^{a} \sigma(x)dx + O(1/z^2)$, we find that the normalisation condition is enforced by

$$\frac{i}{\pi} \int_{-a}^{a} \frac{x^2}{R_+(x)} = 1, \tag{32}$$

which thus gives us a condition to determine a. The integral in Equation (32) is evaluated by elementary means, producing the condition $1 = a^2/2$. Noting that $R_+(z) = -R_-(z)$, the integral in Equation (30) can be rewritten as $F(z) = \frac{R(z)}{2\pi^2} \int_{\mathcal{C}} \frac{\zeta d\zeta}{(\zeta-z)\sqrt{\zeta^2-a^2}}$, where $\mathcal{C}$ is a clockwise contour such that Σ lies inside the contour, and z is outside. Further,

$$F(z) = \frac{R(z)}{2\pi^2} \frac{2\pi iz}{R(z)} + \frac{R(z)}{2\pi^2} \int_{\mathcal{C}_1} \frac{\zeta d\zeta}{(\zeta - z)R(\zeta)}, \tag{33}$$

where $\mathcal{C}_1$ encircles both z and Σ, and the first term cancels the residue of the integrand at z. Since there are no other singularities, $\mathcal{C}_1$ can now be stretched to a circle $\mathcal{C}_2$ of radius $A \to \infty$. Using $\zeta/R(\zeta)(\zeta - z) \to 1/\zeta + O(A^{-2})$ as $|\zeta| = A \to \infty$, the integral is seen to be equal to $\int_{\mathcal{C}_2} \frac{d\zeta}{\zeta} = -2\pi i$. Therefore $F(z) = \frac{i}{\pi}(z - \sqrt{z^2 - a^2})$, and hence, substituting $a = \sqrt{2}$,

$$\sigma(x) = \frac{1}{2}[F_+(x) - F_-(x)] = -\frac{i}{2\pi}\left(2i\sqrt{a^2 - x^2}\right) = \frac{1}{\pi}\sqrt{2 - x^2}. \tag{34}$$

This result is the celebrated *Wigner semicircle law*. In terms of the original variables, $R_1(\lambda) \xrightarrow[N\to\infty]{} \frac{1}{\pi}\sqrt{2N - \lambda^2}$.

The form of the r.h.s. of Equation (24) stems from the Gaussianity of Equation (8). Invariant non-Gaussian ensembles, for example, $P(H) \propto e^{-TrV(H)}$ with some even polynomial $V(x)$, would produce $V'(x)/2$ in the r.h.s. in Equation (24), and therefore generically a different shape of the density of states, as illustrated in the following exercise:

Exercise 4.1. *Given an invariant distribution* $\mathcal{P}(H) \propto e^{-t\mathrm{Tr}H^4}$ *and assuming that the equilibrium measure* $d\mu_0(x)$ *exists and is equal to* $\sigma(x)dx$ *for a continuous function* $\sigma(x)$ *of finite support, show that the support of the density is given by* $x^2 \leq 2/\sqrt{3t}$, *and* $\sigma(x) = \frac{2t}{\pi}\left[x^2 + \frac{1}{\sqrt{3t}}\right]\sqrt{\frac{2}{\sqrt{3t}} - x^2}$.

Revisiting other 'ingredients' of the derivation above, we note that the l.h.s. of Equation (24) is inherited from the Vandermonde determinant in the Jacobian. As the ability to explicitly write down the Jacobian follows from the invariance of Equation (8), the method above is not applicable in general for most non-invariant ensembles. Nevertheless, Wigner's semicircle does possess a certain degree of universality — namely, it obtains for ensembles with *independent* matrix elements under very mild restrictions. Rigorous proof of this fact is outside the scope of these notes [21], however, the sketch of the following alternative method for GUE illustrates why it generalizes to most Wigner ensembles.

Consider the moments of the density of states, $M_{2k} = N^{-k}\mathbb{E}[\mathrm{Tr}H^{2k}] = \int d\mu(x)x^{2k}$, using the counting measure $d\mu(x)$ of rescaled eigenvalues introduced earlier. Under the previous assumptions, if $\mu(x)$ converges to the equilibrium measure, we should have $M_k = \int dx x^{2k}\sigma(x) = 2^{-k}C_k$, where $C_k = \binom{2k}{k}/(k+1)$ are Catalan numbers. On the other hand, evaluating the expectation explicitly, we note that, since matrix elements are uncorrelated and $\mathbb{E}[H_{ij}] = 0$, non-vanishing contributions arise only from combinations of even matchings among the indices in the expansion $\mathrm{Tr}H^{2k} = \sum_{i_1,...i_{2k}} H_{i_1 i_2} H_{i_2 i_3}\ldots H_{i_{2k} i_1}$. For example, using $\mathbb{E}[H_{ij}H_{ji}] = 1/2$, we find for $\mathbb{E}[\mathrm{Tr}H^4] = \mathbb{E}[\sum_{abcd} H_{ab}H_{bc}H_{cd}H_{da}]$ that either $(ab) = (cb)$, giving $\sum_{abd}\mathbb{E}[|H_{ab}|^2|H_{ad}|^2]$, or $(ab) = (ad)$, giving $\sum_{abc}\mathbb{E}[|H_{ab}|^2|H_{bc}|^2]$, or $a = b = c = d$ producing $\sum_a \mathbb{E}[|H_{aa}|^4]$. Since each summation runs from 1 to N, the third contribution is $O(N^{-2})$ compared to the first two contributions. Also, the first two contributions factorize with the exception of the $b = d$ (or $a = c$ in the second case) terms, and the exceptional terms again produce contributions that are lower order in N. It can be shown that at any order the combinatorics of these matchings gives the leading in N contribution that is proportional to the corresponding Catalan number, thus converging to the required moments of $\sigma(x)$ in the $N \to \infty$ limit.

This sketch also illustrates that in the case of more general Wigner distributions, accounting for higher moments of H_{ij} produces lower order contributions, since each time a pair of indices is required to coincide, a summation of $O(N^2)$ terms is removed. This argument can be made rigorous, showing universality of the Wigner semicircle for Wigner ensembles. Moreover, although the sketch above seems to require that all moments of H_{ij} are finite, this restriction can be weakened, so that the density of states has the Wigner semicircle form for all Wigner ensembles with finite second moments of H_{ij} (see Ref. [21]).

Exercise 4.2. *Consider matrices of the form* $\mathbf{M} = \mathbf{H} + v\mathbf{a} \otimes \mathbf{a}^{T*}$, *where* $\mathbf{H}$ *is an* $N \times N$ *matrix drawn from the standard GUE ensemble,* $\mathcal{P}(\mathbf{H}) \propto e^{-\operatorname{Tr}H^2}$, v *is a fixed positive number, and* $\mathbf{a}$ *is a fixed normalised (i.e.,* $\mathbf{a}^{T*}\mathbf{a} = 1$*)* N*-dimensional complex vector. This modification of GUE corresponds to shifting GUE matrices by a rank-1 perturbation (see also Ref. [22]). In the* $N \to \infty$ *limit, find the conditions under which a single eigenvalue splits from the semicircle, and the position of the split eigenvalue.* **Hints:** *(1) Use the identity* $\det(\cdot) = \exp \operatorname{Tr} \ln(\cdot)$ *to establish an equation on the eigenvalues of* $\mathbf{M}$ *in terms of the eigenvalues of* $\mathbf{H}$*, for any finite* N*. (2) Average the resulting equation over* $\mathbf{H}$ *by changing variables to angular integration over the Haar measure of* $U(N)$*, and integration over the eigenvalues of* $\mathbf{H}$*. The following identity is useful here:* $\int d\mu_N(U) U_{ij}(U^\dagger)_{mn} = \frac{1}{N}\delta_{in}\delta_{jm}$, *where* $d\mu_N(U)$ *is the invariant normalised Haar measure on* $U(N)$*, and* U_{ab} *is the ab element of the corresponding* $N \times N$ *unitary matrix. (3) Take the* $N \to \infty$ *limit in the remaining integral over the eigenvalues by using the identity* $\lim_{N\to\infty} \sum_{j=1}^{N} \frac{1}{z - \lambda_j} = \int \frac{R_1(x)dx}{z - x}$*, where in the r.h.s.* $R_1(x)$ *is the semicircle density defined above, the integral is over its support, and* z *lies outside the support of the semicircle.*

4.3. Correlation kernel

Let us now return to the general analysis of correlation functions. The first crucial observation is that, instead of monomials λ_i^j, the Vandermonde determinant can be written in terms of any monic orthogonal polynomials

$p_j(\lambda) = \lambda^j + \sum_{k=0}^{j-1} a_{jk} \lambda^k$. Indeed, by adding to the second row of λ_i^{j-1} the first row multiplied by a_{10}, then adding to the third row the sum of a_{20} times the first row and a_{21} times the (original) second row, *etc.*, we see that $\det \lambda_i^{j-1} = \det p_{j-1}(\lambda_i)$. The freedom to choose *any* polynomials to represent the Vandermonde determinant affords us an opportunity to make a particularly convenient choice: polynomials orthogonal with respect to the weight $e^{-\lambda^2}$, i.e., Hermite polynomials. We can also see that the method is not restricted to Gaussian ensembles — any invariant distribution of the form $e^{-\mathrm{Tr}V(H)}$, where $V(x)$ is an even polynomial with a positive senior coefficient can be accommodated by the appropriate choice of the orthogonal polynomial set.

The second key to the calculation of the correlation functions is the 'integrating-out' lemma [3]. Suppose one is given an $N \times N$ determinant $D^{(N)} = \det d_{ij}$ where $d_{ij} = f(x_i, x_j)$ for some kernel f and measure $d\mu(x)$ possessing the so-called 'reproducing property': $\int d\mu(x_j) f(x_i, x_j)$ $f(x_j, x_k) = f(x_i, x_k)$. It then follows that $\int d\mu(x_N) D^{(N)} = (I - N + 1)$ $D^{(N-1)}$, where $I = \int d\mu(x) f(x, x)$. The proof can be found in Refs. [3] or [24].

Since we have chosen the polynomials to be monic, the orthogonality relations involve normalising constants: $\int p_j(\lambda) p_k(\lambda) e^{-\lambda^2} d\lambda = c_j c_k \delta_{jk}$. Introducing a set of functions $\phi_j(\lambda) = c_j^{-1} p_j(\lambda) e^{-\lambda^2/2}$, we can rewrite the distribution function as $P(\{\lambda\}) \propto (\det A)^2$, where $A_{ij} = \phi_{i-1}(\lambda_j)$. Since $(\det A)^2 = \det[A^T A]$, where A^T is the transpose of A, we find

$$P(\{\lambda\}) \propto \det_{i,j \leq N} \left(\sum_{k=1}^{N} A_{ki} A_{kj} \right) = \det_{i,j \leq N} K_N(\lambda_i, \lambda_j), \qquad (35)$$

where $K_N(\lambda, \mu) = \sum_{j=0}^{N-1} \phi_j(\lambda) \phi_j(\mu)$. The orthogonality of the polynomials now ensures that K_N is a reproducing kernel, with $I = \sum_{j=0}^{N-1} \int d\lambda \phi_j^2(\lambda) = N$. We can now use the 'integrating-out' lemma to deduce that

$$R_m(\lambda_1, \ldots, \lambda_m) = \det \left(K_N(\lambda_i, \lambda_j) \right)_{1 \leq i,j \leq m}. \qquad (36)$$

The overall coefficient of Equation (36) is fixed by the normalisation of the correlation functions: from the definition Equation (21) it is obvious that $\int \prod_{j=1}^{m} d\lambda_j R_m(\lambda_1, \ldots, \lambda_m) = N!/(N-m)!$, and the same normalisation of Equation (36) is verified by repeated application of the 'integrating-out' lemma. In particular,

$$R_1(\lambda) = K_N(\lambda, \lambda) = \sum_{j=0}^{N-1} \phi_j(\lambda)^2, \qquad (37)$$

and

$$R_2(\lambda, \mu) = K_N(\lambda, \lambda)K_N(\mu, \mu) - K_N(\lambda, \mu)K_N(\mu, \lambda). \tag{38}$$

4.4. *Orthogonal polynomials approach*

The task of evaluating the correlation functions in the $N \to \infty$ limit is thus transformed into the task of evaluating the asymptotic properties of orthogonal polynomials. Our consideration here is limited to Hermite polynomials — those associated with the Gaussian weight on $\mathbb{R}$, but these techniques can be transferred to a wider class of classical orthogonal polynomials, and hence to the corresponding invariant distribution functions.[b]

From now on we switch to scaled eigenvalues $\epsilon_j = \lambda_j/\sqrt{N}$. The normalised Hermite polynomials are defined as

$$p_n(x) = \frac{1}{\left(2^n N^{n-1/2}\sqrt{\pi}n!\right)^{1/2}} h_n(x), \tag{39}$$

where

$$h_n(x) = (-1)^n e^{Nx^2} \frac{d^n}{dx^n} e^{-Nx^2}. \tag{40}$$

A general property of orthogonal polynomials is the existence of three-term recurrence relations, taking the following form in this case:

$$\sqrt{\frac{n+1}{2N}} p_{n+1}(x) = x p_n(x) - \sqrt{\frac{n}{2N}} p_{n-1}(x). \tag{41}$$

Equation (41) can be used to derive the following Christoffel–Darboux formula

$$\sum_{j=0}^{n-1} p_j(x)p_j(y) = \sqrt{\frac{n}{2N}} \frac{p_{n-1}(y)p_n(x) - p_{n-1}(x)p_n(y)}{x - y}. \tag{42}$$

Taking the limit $y \to x$, we also find its confluent form

$$\sum_{j=0}^{n-1} p_j^2(x) = \sqrt{\frac{n}{2N}} [p_{n-1}(x)p_n'(x) - p_n(x)p_{n-1}'(x)]. \tag{43}$$

Note that according to Equations (37) and (38), we are interested in the $N \to \infty$ behaviour of p_N and p_{N-1}, thus the $N \to \infty$ limit affects

---

[b]This section owes significantly to the presentation in Ref. [32].

simultaneously the argument and the index dependence of the polynomials. Such behaviour is controlled by the so-called Plancherel–Rotach asymptotics of the orthogonal polynomials. The latter can be obtained by applying the steepest descent method to the integral representation

$$h_n(x) = (-2iN)^n \sqrt{N} \pi e^{Nx^2} \int_{-\infty}^{\infty} dz z^n e^{-Nz^2 + 2ixzN} \qquad (44)$$

of the polynomials. Omitting the details [32], the asymptotics of the polynomials takes the form

$$p_{N+m}(x) \longrightarrow e^{Nx^2/2} \sqrt{\frac{\sqrt{2}}{\pi \sin \phi}} \cos \left[\left(N + \frac{1}{2} \right) \phi + m\phi - \frac{\pi}{4} - \frac{N}{2} \sin 2\phi \right],$$
$$(45)$$

where $(1 + \cos 2\phi)/2 = x^2/2$.

In order to apply the asymptotic result in Equation (45) to the calculation of R_1, it is convenient to transform Equation (43) (evaluated at $n = N$) using the following relation (easily inferred from Equation (44): $h'_k(x) = 2Nxh_k(x) - h_{k+1}(x)$. This relation implies in the leading $N \to \infty$ approximation

$$p'_N(x)p_{N-1}(x) - p'_{N-1}(x)p_N(x) \longrightarrow \sqrt{2}N \left(p_N^2(x) - p_{N-1}(x)p_{N+1}(x) \right). \qquad (46)$$

We thus find

$$R_1 = e^{-Nx^2} \sqrt{\frac{N}{2N}} \left[p'_N(x)p_{N-1}(x) - p'_{N-1}(x)p_N(x) \right]$$

$$\approx \frac{\sqrt{2}N}{\pi \sin \phi} \left[\cos^2 A - \cos(A - \phi) \cos(A + \phi) \right], \qquad (47)$$

where $A = \left(N + \frac{1}{2} \right) \phi - \frac{\pi}{4} - \frac{N}{2} \sin 2\phi$. Using $\cos^2 A - \cos(A - \phi) \cos(A + \phi) = \sin^2 \phi$, we obtain

$$R_1 \approx \frac{N\sqrt{2}}{\pi} \sin \phi = \frac{N}{\pi} \sqrt{2 - x^2} = N\sigma(x), \qquad (48)$$

reproducing, as expected, the Wigner semicircle result.

4.4.1. *Correlation kernel*

A similar calculation for the asymptotic form of the kernel $K_N(x,y)$ gives

$$K_N = \frac{1}{\sqrt{2}} e^{-\frac{N}{2}(x^2+y^2)} \frac{p_N(x)p_{N-1}(y) - p_N(y)p_{N-1}(x)}{x-y}$$

$$\approx \frac{1}{\pi\sqrt{\sin\phi(x)\sin\phi(y)}}$$

$$\times \frac{\cos A(x)\cos(A(y)-\phi(y)) - \cos A(y)\cos(A(x)-\phi(x))}{x-y}, \quad (49)$$

where $A(\xi) = A|_{\phi=\phi(\xi)}$, and $\sqrt{2}\cos\phi(\xi) = \xi$. This expression can be further transformed in the so-called *bulk scaling limit*, i.e., under the assumption that $|x-y| \sim O(N^{-1})$ and that both x and y lie inside the support of the spectrum $(-\sqrt{2}, \sqrt{2})$. Expanding $\phi(x) \approx \phi(y) + \phi'(y)(x-y)$, and using $\phi'(y) = 1/\pi\sigma(y)$, we find

$$A(x) \approx A(y) + \frac{N(x-y)}{\pi\sigma(y)}[1 - \cos 2\phi(y)] = A(y) + N\pi\sigma(y)(x-y), \quad (50)$$

where we have used $1 - \cos 2\phi(y) = 2\sin^2\phi(y) = \pi^2\sigma(y)^2$. Note that when performing the expansion in Equation (49) we consistently keep the terms of order N and of order 1 under the cosines, while neglecting the higher-order terms. Denoting $\omega = N\sigma(y)(x-y)$, we use

$$\cos(A(y)+\pi\omega)\cos(A(y)-\phi(y)) - \cos A(y)\cos(A(y)+\pi\omega-\phi) = \sin\phi(y)\sin\pi\omega$$
$$(51)$$

to establish finally the following limiting value of the kernel $K_N(x,y)$ in the scaling limit $|x-y| \sim O(N^{-1})$:

$$\lim_{N\to\infty} K_N\left(x, x+\frac{\omega}{N\sigma(x)}\right) = N\sigma(x)\frac{\sin\pi\omega}{\pi\omega} \equiv N\sigma(x)\mathcal{K}(\omega). \quad (52)$$

This is the famous *sine kernel* of the Wigner–Dyson theory, giving the universal behaviour of correlation functions in the $N \to \infty$ limit.

The scaling limit Equation (52) implies that the two-point correlation function (38) acquires, in the bulk scaling limit, the form

$$R_2\left(x, x+\frac{\omega}{N\sigma(x)}\right) = N^2\sigma^2(x)\mathcal{R}_2(\omega) = N^2\sigma^2(x)\left[1 - \frac{\sin^2\pi\omega}{\pi^2\omega^2}\right], \quad (53)$$

provided x lies inside the support of the spectrum $(-\sqrt{2}, \sqrt{2})$ and $\omega \sim O(1)$. Note that $\Delta = 1/N\sigma(\epsilon)$ (not to be confused with the Vandermonde

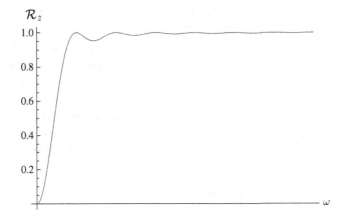

Fig. 2. The two-point correlation function in the unitary ensemble.

determinant) has the meaning of the mean eigenvalue spacing. Thus, the universality of the correlation functions in the scaling limit can be restated as the property that the eigenvalue correlations have universal functional forms when restricted to eigenvalues separated by 'distances' of the order of mean eigenvalue separation. As mentioned in the Introduction, the existence of the specific universal functional form of the kernel controlling the correlation functions is conceptually similar to the Central Limit Theorem. The determinantal form of general correlation functions R_m ensures that they inherit the universal form from the universality of the sine kernel.

The graph of the two-point function is presented in Fig. 2. In many applications, a convenient object to study is its Fourier transform (the so-called *form-factor*), which has a remarkably simple structure:

$$\tilde{\mathcal{R}}_2(t) \equiv \int_{-\infty}^{\infty} d\omega \left[\mathcal{R}_2(\omega) - 1 + \delta(\omega)\right] e^{2\pi i \omega t} = |t|\theta(1 - |t|) + 1 \cdot \theta(|t| - 1),$$

(54)

where the term $\delta(\omega)$ is added to account for the 'self-correlation' of eigenvalues (the two eigenvalues in the derivation leading to Equation (38) were always assumed to be distinct), and 1 is subtracted to eliminate the trivial effect of the lack of correlations at large separations.

4.4.2. Universality

The derivation above, based as it is on the intricate asymptotic analysis of Hermite polynomials, somewhat obscures the origins of the most fundamental fact about the sine kernel — its universality. In fact, more advanced methods can be used to show that universality of the sine kernel in the bulk scaling limit extends for a broad class of ensembles of Hermitian matrices. For the discussion of invariant non-Gaussian ensembles of the $P(H) \propto e^{-\mathrm{Tr}V(H)}$ type, including non-polynomial $V(H)$, see Refs. [24, 33]. A way to understand the 'mechanics' of universality in invariant ensembles is through the observation that asymptotic behaviour of orthogonal polynomials with smooth weights at large index value always incorporates oscillations, since frequency mismatch between oscillating functions is the primary 'mechanism' to achieve orthogonality.

Exercise 4.3. *An important class of ensembles as yet not mentioned is the* circular ensembles *— ensembles of matrices from the classical groups, distributed according to the Haar measure [3]. One advantage in using these ensembles is that the (now complex) eigenvalues have uniform density on the unit circle. Another feature of these ensembles is that the kernel is expressed in terms of simple trigonometric functions for any N rather than asymptotically for large N. In the case of unitary group* $\mathrm{U}(N)$*, the joint distribution of the eigenvalues* $\{\theta_i\}_{i=1}^N$ *can be shown to be proportional to the squared Vandermonde determinant built on the corresponding phases:* $\prod_{i<j}|e^{i\theta_i}-e^{i\theta_j}|^2$*. Use this expression to show that the analogue of the kernel* $K_N(x,y)$ *is* $S_N(\theta,\chi)$*, where* $S_N(\theta) = (1/2\pi)\sin(N\theta/2)/\sin\theta/2$*, and hence show that the two-point correlation function is given by the same universal expression* $\mathcal{R}_2$ *in the* $N \to \infty$ *limit.*

The possibilities for non-invariant ensembles are, of course, nearly limitless, and not all of them exhibit this universality. The study of universality in ensembles without invariance is technically more challenging, essentially because the intermediate step of deriving the joint distribution of all $O(N)$ eigenvalues is unavailable. Although much has been known for several decades at the 'physical level of rigor' [25], substantial progress towards rigorous proof of universality for Wigner ensembles has only happened in the last decade [23]. The key idea behind these proofs relies on Dyson diffusion starting from a Wigner matrix, and using an argument that such Dyson-evolved ensembles with a small normal component are dense in the space of Wigner matrices. Some other examples of non-invariant ensembles for which the validity of the sine kernel in the bulk scaling limit was shown include distributions of the form $e^{\mathrm{Tr}V(H)+\mathrm{Tr}[AH]}$, where A is a constant matrix (Refs. [34] or [35] for a more formal proof), and $e^{\mathrm{Tr}V(H-A)}$ (Ref. [37] at the 'physical level of rigour', or Ref. [22] for a recent more formal consideration).

(Very) qualitatively, the guiding principle that emerged from these studies is that universal behaviour of spectral correlations emerges when there is a sufficient number and strength of non-vanishing off-diagonal matrix elements to ensure that the original (before diagonalisation) basis states are effectively coupled. An essentially equivalent observation is that spectral correlations are governed by the universal sine kernel whenever eigenfunctions are *delocalized* — i.e., roughly, whenever the eigenstates generically contain $O(N)$ non-vanishing (more precisely, not exponentially small) components in the original basis. [It can be seen that invariant ensembles satisfy this condition trivially, as the eigenstates are rows of a unitary matrix drawn from an invariant Haar measure.] A useful case in point is the study of sparse random matrices [36], where each row contains $O(1)$ nonzero entries. It turns out that there exists a critical value $p_c \sim O(1)$ such that the eigenstates are delocalized and the spectral statistics is universal if the probability for a matrix element to be nonzero is greater than p_c/N. Although the derivation in this section was confined to the technically simplest case $\beta = 2$, the review of universality above, with some exceptions, applies to all three values of β, each of which is characterised by its own universal kernel.

4.4.3. *Number variance*

A very informative characteristic of the eigenvalue sequence is given by the so-called *number variance*, defined as

$$\Sigma(S) = \mathbb{E}[N(S)^2] - \left(\mathbb{E}[N(S)]\right)^2, \tag{55}$$

where $N(S)$ is the number of eigenvalues in an interval of length S. For example, as is well known, a sequence drawn from a Poisson process of intensity $1/\Delta_P$ would have $\mathfrak{n} = N(S)$ distributed according to $P(\mathfrak{n}) = \frac{(S/\Delta_P)^{\mathfrak{n}}}{\mathfrak{n}!}e^{-S/\Delta_P}$, and therefore $\Sigma_P(S) = S/\Delta_P$, growing linearly with S. In contrast, an eigenvalue sequence is characterised, for $S \sim O(\Delta)$, by

$$\Sigma(S) = \int_{\epsilon_0}^{\epsilon_0+S} d\epsilon[N\sigma(\epsilon)]$$

$$-N^2 \int_{\epsilon_0}^{\epsilon_0+S} d\epsilon \int_{\epsilon_0}^{\epsilon_0+S} d\epsilon'\sigma(\epsilon)\sigma(\epsilon')\mathcal{K}^2\left[N\sigma(\epsilon)(\epsilon-\epsilon')\right]$$

$$= s - \int_0^s dx \int_0^s dy\mathcal{K}^2(x-y), \tag{56}$$

where $s = S/\Delta = N\sigma(\epsilon_0)S$, and the first term is included in order to compensate for the fact that the correlation function $R_2(\epsilon, \epsilon')$ is defined in such a way that ϵ and ϵ' always refer to distinct eigenvalues. Skipping the technical details [3], in the asymptotic limit $s \gg 1$ the leading contribution to the number variance is

$$\Sigma(S) = \frac{1}{\pi^2}[\ln 2\pi s + \gamma + 1] + O\left(s^{-1}\right). \tag{57}$$

As compared to the Poisson process, the growth of the number variance is considerably slower — logarithmic as opposed to linear. This phenomenon is another consequence of eigenvalue repulsion: eigenvalues mutually 'hem each other in' so that the overall sequence is considerably more rigid than an uncorrelated Poissonian sequence, with correlations decaying algebraically. Another manifestation of the same phenomenon is the fact that the oscillating factor in $\mathcal{K}$ has the period equal to the mean eigenvalue spacing, thus retaining an algebraically decaying 'remnant' of the information on the expected eigenvalue positions at arbitrarily large separations $|x - y| \gg 1$.

4.5. *Gap probability and eigenvalue spacing distribution*

In principle, the knowledge of *all* m-point correlation functions enables us to determine another key quantity often used in analyzing spectral sequences: the so-called *gap probability* $\mathcal{A}(S)$, defined as the probability that the interval $\mathcal{S} = (0, S)$ contains no eigenvalues. Formally, introducing the characteristic function $\chi_B(x)$ of a set B via $\chi_B(x) = 1$ if $x \in B$, and $\chi_B(x) = 0$ otherwise, we write

$$\mathcal{A}(S) = \int \prod_{i=1}^{N}(1 - \chi_{\mathcal{S}}(\lambda_i))P(\{\lambda\}). \tag{58}$$

Expanding the product of $(1 - \chi_{\mathcal{S}})$ we find

$$\mathcal{A}(S) = \sum_{m=0}^{N} \frac{(-1)^m}{m!} \int_0^S \prod_{j=1}^{m} d\lambda_j R_m(\lambda_1, \dots, \lambda_m). \tag{59}$$

Therefore

$$\mathcal{A}(S) = \sum_{j=0}^{N} \frac{(-1)^j}{j!} \int_0^S \prod_{i=1}^{j} d\lambda_i \det\left(K_N(\lambda_k, \lambda_l)\right)_{1 \leq k,l \leq m} = \det_{[0,S]}\left(\mathbb{I} - \mathbb{K}_N\right), \tag{60}$$

where $\mathbb{K}_N$ is the operator with the kernel $K_N(\lambda, \mu)$. This operator has finite rank, and is therefore, trace class.

Directly inherited from the sine kernel is the universal $N \to \infty$ formula for the gap probability: in terms of the scaled variable $s = S/\Delta$, the probability that an interval of length s is free of eigenvalues is given by the Fredholm determinant

$$\mathcal{A}(s) = \det_{[0,s]}(1 - \hat{\mathcal{K}}), \tag{61}$$

where the kernel of the integral operator $\hat{\mathcal{K}}$ is the sine kernel. The nearest-neighbour spacing distribution $p(s)$ is given by $\frac{d^2}{ds^2}\mathcal{A}(s)$ [3].

4.5.1. *Wigner surmise*

The Fredholm determinant in Equation (61) cannot be evaluated in closed form. An elementary approximation to eigenvalue spacing distributions can be obtained from the 2×2 'toy' models. Integrating the joint distributions (2) and (3) over the midpoint $(\lambda_1 + \lambda_2)/2$, we obtain that the distributions of spacings $S = |\lambda_1 - \lambda_2|$ for real symmetric and complex 2×2 matrices are $P_1(S) = \frac{S}{4}e^{-S^2/4}$ in the real case, and $P_2(S) = \frac{2S^2}{\sqrt{2\pi}}e^{-S^2/2}$ in the complex Hermitian case. Introducing the scaled variable $s = S/\mathbb{E}[S]$ we find (easily computing $\mathbb{E}[S] = \sqrt{\pi}$ and $2\sqrt{2/\pi}$ in the two cases)

$$p_1(s) = \frac{\pi s}{2}e^{-\pi s^2/4}, \quad p_2(s) = \frac{32s^2}{\pi^2}e^{-4s^2/\pi}. \tag{62}$$

This is the famous *Wigner surmise* [3] for the distribution of eigenvalue spacings. A similar expression can be derived for the symplectic case. Although it was obtained here through the use of 2×2 'toy' models, this expression turns out to be a remarkably good approximation to the actual (universal!) distribution of nearest neighbour eigenvalue spacings of large matrices.

4.5.2. *Asymptotics of the eigenvalue spacing distribution*

The Gaussian decay at large s in the above derivation — much faster than the exponential decay of the corresponding quantity in a Poissonian sequence — was inherited from the assumed distribution function. Such a direct 'inheritance', i.e., explicit dependence on the specific form of the distribution function of the matrix elements, runs counter to the RMT 'ideology' of universality. In fact, the true asymptotic decay at large s [given by the asymptotics of (61) in the unitary case] does happen to be Gaussian for the same reason that the number variance is logarithmic: the collection of eigenvalues is very rigid, so that creating a large gap requires 'pushing out' a large number of fairly distant eigenvalues.

The true asymptotic behaviour as $s \to \infty$ in the unitary case is given by

$$\ln \mathcal{A}(s) \xrightarrow[s \to \infty]{} -\pi^2 s^2/8 - (1/4)\ln s + \text{Const.} + O(1/s). \tag{63}$$

Note that the exact coefficient of the leading quadratic term is slightly different from the Wigner surmize: $\pi^2/8 \approx 1.2337$, while $4/\pi \approx 1.2732$. Although the leading coefficient can be obtained by fairly elementary means, for example by minimising Equation (23) subject to the additional constraint that the measure vanishes on the $(0, s)$ interval, calculation of the remaining terms requires more advanced approaches, such as solving *matrix* Riemann–Hilbert problems (see Ref. [24] for a detailed discussion and further references).

4.6.   *A sketch of Pastur–Scherbina method*

An alternative to the Plancherel–Rotach asymptotics route for deriving the $N \to \infty$ asymptotics of the kernel is given by Pastur and Shcherbina [33]. The main disadvantage of the Plancherel–Rotach scheme is that it requires the knowledge of the integral representation of the relevant orthogonal polynomials. If the distribution is invariant, but not Gaussian, i.e., $\mathcal{P}(H) \propto e^{-N\mathrm{Tr}V(H)}$ for some function $V(H)$ which grows fast enough to ensure normalisablity of the distribution,[c] the integral representation of the corresponding polynomials orthogonal with respect to the weight $e^{-V(x)}$ is available only in the special cases when $V(x)$ is the weight function of one of the classical orthogonal polynomials. The approach by Pastur and Scherbina [33] is free from this constraint. The starting point of this calculation is the following representation of the kernel:

$$K_N(\epsilon, \epsilon') = \frac{1}{(N-1)!} e^{-(N/2)[V(\epsilon)+V(\epsilon')]} \int \prod_{j=2}^{N} d\epsilon_j e^{-N\sum_{j=2}^{N} V(\epsilon_j)}$$

$$\times \prod_{j=2}^{N} (\epsilon - \epsilon_j)(\epsilon' - \epsilon_j)\Delta^2\left(\{\epsilon_j\}_{j=2}^{N}\right). \tag{64}$$

The validity of Equation (64) can be proved analogously to the 'integrating-out' lemma.

---

[c]In a slight change of notation, here it is convenient to scale the distribution with N.

It is assumed that $V(x) > 2(1 + \varepsilon)\ln(1 + |x|)$ for some constant ε, and some additional technical restrictions on $V(x)$ are imposed. Now, anticipating the scaling limit, let us denote $\mathcal{K}_N(x,y) = (1/N)K_N(\epsilon_0 + x/N, \epsilon_0 + y/N)$. Differentiating Equation (64), one obtains

$$\frac{\partial}{\partial x}\mathcal{K}_N(x,y) = -\frac{1}{2}V'(\epsilon_0 + x/N)\mathcal{K}_N(x,y)$$

$$+ \int dz \frac{\mathcal{K}_N(z,z)\mathcal{K}_N(x,y) - \mathcal{K}_N(x,z)\mathcal{K}_N(z,y)}{x-z}. \quad (65)$$

Guided by Equation (24), it is reasonable to assume that

$$D(\epsilon) = \frac{1}{2}V'(\epsilon) - \frac{1}{N}\int \frac{\mathcal{K}_N(\mu,\mu)}{\epsilon - \mu} \quad (66)$$

must vanish in the $N \to \infty$ limit. Indeed [33], $D(\epsilon) \sim o(N)$, and one obtains the following integral equation on $\mathcal{K}_N$:

$$\frac{\partial}{\partial x}\mathcal{K}_N(x,y) = -\int dz \frac{\mathcal{K}_N(x,z)\mathcal{K}_N(z,y)}{x-z} + r(N), \quad (67)$$

where $r(N)$ is the error term which vanishes in the $N \to \infty$ limit

The key element of the proof is a series of estimates which show that $\mathcal{K}_N(x,y)$ converges to a function $\mathcal{K}^*(x-y)$ of the difference only. The latter, therefore, satisfies the integral equation

$$\frac{\partial}{\partial x}\mathcal{K}^*(x) = -\int dz \frac{\mathcal{K}^*(z)\mathcal{K}^*(x-z)}{z}. \quad (68)$$

Introducing $F(p) = \int_0^p dp' \int dx e^{ixp'}\mathcal{K}^*(p')$ one obtains $\int (F(p) - p)h(p)dF$ $(p) = 0$ for any smooth function $h(p)$ of compact support, which has to be satisfied subject to $F(\infty) - F(-\infty) = 2\pi\sigma(\epsilon_0)$, and some technical restrictions which ensure that the solution

$$F(p) = p\theta(p_0 - |p|) + \mathrm{Sgn}p\theta(|p| - p_0), \quad (69)$$

with $p_0 = \pi\sigma(\epsilon_0)$ is unique. The constraint on F which establishes the scale of p_0 stems from the fact that $\mathcal{K}_N$ at coinciding arguments converges to the eigenvalue density. Fourier transform of Equation (69) gives the sine kernel.

What is truly remarkable about this proof is the fact that the sine kernel here appears as a solution of an integral equation whose structure is directly inherited from the structure of the Vandermonde determinant. In this sense, the sine kernel here emerges as an immediate consequence of the core properties of the eigenvalue sequence as compared to the alternative approach where it emerges at some remove as an asymptotic property of orthogonal polynomials, and thus may *a priori* involve the properties of the weight function defining the polynomials.

References

[1] J. Wishart, *Biometrika A*, **20** (1928), 32.
[2] E. P. Wigner, *Proc. Cambridge Phil. Soc.*, **47** (1951), 790.
[3] M. L. Mehta, *Random Matrices*, 3rd ed., Elsevier, Amsterdam, 2004.
[4] M. Gaudin, *Nuclear Phys.*, **25** (1961), 447.
[5] F. J. Dyson, *J. Math. Phys.*, **3** (1962), 140; 157; 166.
[6] F. J. Dyson, *J. Math. Phys.*, **3** (1962), p. 1191.
[7] F. J. Dyson, *J. Math. Phys.*, **3** (1962), p. 1199.
[8] O. Bohigas, M. J. Giannoni and C. Schmit, *Phys. Rev. Lett.*, **52** (1984), 1.
[9] G. Casati and F. Vals-Griz, *Lett. Nuovo Cimento*, **28** (1980), 279.
[10] T. Guhr, A. Mueller-Groeling and H. Weidenmueller, *Phys. Rep.*, **299** (1998), 189.
[11] M. Giesen and T. L. Einstein, *Surf. Sci.*, **449** (2000), 191.
[12] M. Prahofer and H. Spohn, *Phys. Rev. Lett.*, **84** (2000), 4882.
[13] G. Akemann, J. Baik, P. Di Francesco, eds., *The Oxford Handbook of Random Matrix Theory*, Oxford University Press, Oxford, 2011.
[14] V. Plerou *et al.*, *Phys. Rev. E*, **65** (2002), 066126.
[15] P. Seba, *Phys. Rev. Lett.*, **91** (2003), 198104.
[16] A. Tulino and S. Verdú, *Foundations and Trends® in Communications and Information Theory*, **1** (2004), 1.
[17] H. Montgomery, *Proc. Sympos. Pure Math*, **24** (1973), 181.
[18] A. Odlyzko, *Math. Comput.*, **48** (1987), 273.
[19] A. Connes, *CR Acad. Sci. I-Math*, **323** (1996), 1231.
[20] G. Sierra, *J. Phys. A*, **47** (2014), 325204.
[21] G. Anderson, A. Guionnet and O. Zeitouni, *An Introduction to Random Matrices*, Cambridge University Press, Cambridge, 2010.
[22] P. J. Forrester, *Log-gases and random matrices*, Princeton University Press, Princeton, 2010.
[23] T. Tao, *Topics in Random Matrix Theory*, Graduate Studies in Mathematics, 132. American Mathematical Society, Providence, RI, 2012.
[24] P. Deift, *Orthogonal Polynomials and Random Matrices*, AMS Providence, RI, 2000.
[25] K. Efetov, *Supersymmetry in Disorder and Chaos*, Cambridge University Press, Cambridge, 1997.
[26] J. Verbaarschot, H. Weidenmüller and M. Zirnbauer, *Phys. Rep.* **129** (1985), 367.
[27] D. Voiculescu, *Invent. Math.*, **104** (1991), 201.
[28] F. Haake, *Quantum Signatures of Chaos*, Springer, Berlin, 2001.
[29] G. Grimmett and D. Stirzaker, *Probability and Random Processes*, 3rd ed., Oxford University Press, Oxford, 2001.
[30] R. Tribe and O. Zaboronski, *J. Math. Phys.*, **55** (2014), 063304.
[31] N. Muskhelishvili, *Singular Integral Equations*, Dover Publications, New York, 1992.
[32] Y. V. Fyodorov, *preprint* arXiv:math-ph/0412017.

[33] L. Pastur and M. Shcherbina, *J. Stat. Phys.*, **130** (2008), 205.

[34] P. Zinn-Justin, *Nucl. Phys. B*, **497** (1997), 725; *Comm. Math. Phys.*, **194** (1998), 631.

[35] P. Bleher and A. Kuijlaars, *Comm. Math. Phys.*, **252** (2004), 43; *ibid.*, **270** (2007), 481; A. Aptekarev, P. Bleher and A. Kuijlaars, *ibid.*, **259** (2005), 367.

[36] A. Mirlin and Y. Fyodorov, *J. Phys. A*, **24** (1991), 2273; R. Kühn, *J. Phys. A*, **41** (2008), 295002; J. Huang, B. Landon and H.-T. Yau, *J. Math. Phys.* **56** (2015), 123301.

[37] I. Smolyarenko and B. Simons, *J. Phys. A*, **36** (2003), 3551.

Chapter 6

Symmetry Methods
for Differential Equations

Peter A. Clarkson

School of Mathematics, Statistics & Actuarial Science
University of Kent, Canterbury, CT2 7NF, UK
P.A.Clarkson@kent.ac.uk

In this chapter, we discuss applications of symmetry analysis (sometimes called similarity analysis) to find symmetry reductions and exact analytical solutions of nonlinear partial differential equation.

1. Introduction

Nonlinear phenomena have many important applications in several aspects of physics (as well as other natural and applied sciences). Essentially all the fundamental equations of physics are nonlinear and, in general, such nonlinear equations are often very difficult to solve explicitly. Consequently perturbation, asymptotic and numerical methods are often used (with much success) to obtain *approximate* solutions of these equations; however, there is also much current interest in obtaining *exact* analytical solutions of nonlinear equations. Symmetry group techniques provide one method for obtaining a class of special solutions and furthermore they do not depend upon whether or not the equation is "integrable" or "solvable" (in any sense of the words).

Symmetry groups have several different applications in the context of nonlinear differential equations:

- *Derive new solutions from old solutions.* Applying the symmetry group of a differential equation to a known solution yields a family of new solutions (quite often interesting solutions can be obtained from trivial ones).

- *Integration of ordinary differential equations.* Symmetry groups of ordinary differential equations can be used to reduce the order of the equation (such as to reduce a second-order equation to first-order one).
- *Reductions of partial differential equations.* Symmetry groups of partial differential equations are used to reduce the total number of dependent and independent variables (e.g., reduce a partial differential equation with two independent and one dependent variables to an ordinary differential equation).
- *Linearisation of partial differential equations.* Symmetry groups can be used to discover whether or not a partial differential equation can be linearised and to construct an explicit linearisation when one exists.
- *Classification of equations.* Symmetry groups can be used to classify differential equations into equivalence classes and choose key representatives of such classes.
- *Asymptotics of solutions of partial differential equations.* It is known that as solutions of partial differential equations asymptotically tend to solutions of lower-dimensional equations obtained by symmetry reduction, and some of these special solutions illustrate important physical phenomena. In particular, exact solutions arising from symmetry methods can often be effectively used to study properties such as long-time asymptotics and "blow-up".
- *Numerical methods and testing computer coding.* Symmetry groups and exact solutions of physically relevant partial differential equations are used in the design, testing and evaluation of numerical algorithms; these solutions provide an important practical check on the accuracy and reliability of such integrators.
- *Conservation Laws.* The application of symmetries to conservation laws dates back to the work of Noether who proved the remarkable result that for systems arising from a variational principle, every conservation law of the system comes from a corresponding symmetry property.
- *Further Applications.* There are several other important applications of symmetry groups including bifurcation theory, control theory, special function theory, boundary value problems and free boundary problems.

In the mid-19th century, Sophus Lie was searching for a general theory for solving differential equations. He made the profound and far-reaching discovery that the special methods (such as separable equations, homogeneous equations, exact equations and integrating factors, which until then had seemed not intrinsically related to each other) for solving first-order ordinary differential equations, were, in fact, all special cases of a general integration method based on the invariance of differential equations under a continuous group of symmetries, now known as a *Lie group* (a group which depends upon a continuous parameter). Lie developed a theory for symmetry groups of differential equations which is highly algorithmic. Moreover, Lie provided a classification of all ordinary differential equations of arbitrary order in terms of their symmetry groups, and thus described the whole set of equations integrable by group-theoretical methods. This method is now commonly referred to as the "classical Lie method for finding group-invariant solutions" which was first described in full generality by Lie [1]. For a modern description see, for example, the books by Arrigo [2], Bluman and Kumei [3], Cantwell [4], Hill [5], Hydon [6] and Olver [7].

Subsequently there have been many applications in numerous areas of Mathematics, Physics, Chemistry, Engineering, etc. However, a major difficulty in applying the theory to differential equations is that the task of finding the symmetry group of a given system of differential equations is often exceedingly cumbersome. Despite the fact that the method is entirely algorithmic, it usually involves a large amount of tedious algebra and the associated calculations (sometimes involving hundreds or even thousands of equations), can be virtually unmanageable if attempted manually. However, symbolic manipulations programs have been developed, e.g., in MAPLE [8] and Mathematica [4], to facilitate these calculations.

Our primary application will be the use of symmetry groups to reduce a given partial differential equation to an ordinary differential equation. However, it is still necessary to solve the ordinary differential equation in order to obtain exact solutions.

2. One-parameter groups of transformations

First, we define a one-parameter group of transformations. For simplicity we initially restrict ourselves to two variables x and y.

Definition 1. In the (x, y) plane, the transformation

$$x^* = f(x, y; \varepsilon), \qquad y^* = g(x, y; \varepsilon), \tag{1}$$

is a *one-parameter group of transformations* $\mathcal{G}$ with parameter $\varepsilon \in \mathbb{C}$, if the following properties hold:

(i) the value $\varepsilon = 0$ characterises the *identity* transformation, i.e.,

$$x = f(x, y; 0), \quad y = g(x, y; 0);$$

(ii) the parameter $-\varepsilon$ characterises the *inverse* transformation, i.e.,

$$x = f(x^*, y^*; -\varepsilon), \quad y = g(x^*, y^*; -\varepsilon);$$

(iii) if $x^{**} = f(x^*, y^*; \delta)$, $y^{**} = g(x^*, y^*; \delta)$, then the product (composition) of the two transformations is also a member of the set of transformations (1) characterised by the parameter $\varepsilon + \delta$, i.e.,

$$x^{**} = f(x, y; \varepsilon + \delta), \quad y^{**} = g(x, y; \varepsilon + \delta),$$

so there is closure under composition; and

(iv) the associativity law for groups holds; this property follows from the formula given in (iii) since addition in $\mathbb{C}$ is associative.

Example 1. Examples of one-parameter groups of transformations are given by:

$$x^* = x + \alpha\varepsilon, \qquad\qquad y^* = y + \beta\varepsilon, \qquad\qquad \text{(2a)}$$
$$x^* = x\,e^{\alpha\varepsilon}, \qquad\qquad y^* = y\,e^{\beta\varepsilon}, \qquad\qquad \text{(2b)}$$
$$x^* = x\cos\varepsilon - y\sin\varepsilon, \qquad y^* = x\sin\varepsilon + y\cos\varepsilon, \qquad \text{(2c)}$$
$$x^* = \frac{x}{1 - \varepsilon x}, \qquad\qquad y^* = \frac{y}{1 - \varepsilon x}, \qquad\qquad \text{(2d)}$$

where ε is the group parameter and α and β are arbitrary parameters. The transformation defined by (2a) is referred to as a *group of translations*; the transformation (2b) is a *group of scalings* (sometimes called a *group of dilations*); the transformation (2c) is a *group of rotations*; and the transformation (2d), is a *projective group*. Each of the one-parameter group of transformations (2) have invariants associated with them. The invariants associated transformations (2) respectively are

$$x - y = x^* - y^*, \qquad x^2 + y^2 = (x^*)^2 + (y^*)^2,$$
$$\frac{y^\alpha}{x^\beta} = \frac{(y^*)^\alpha}{(x^*)^\beta}, \qquad \frac{xy}{x - y} = \frac{x^* y^*}{x^* - y^*}.$$

Example 2. Show that the rotation group (2c) forms a one-parameter group of transformations as defined above.

For the rotation group (2c) we need to verify that the properties (i)–(iv) in Definition 1.1 are satisfied.

(1) Clearly property (i) is satisfied since if $x^*|_{\varepsilon=0} = x$ and $y^*|_{\varepsilon=0} = y$.

(2) Solving

$$x^* = x \cos \varepsilon - y \sin \varepsilon, \qquad y^* = x \sin \varepsilon + y \cos \varepsilon,$$

for x and y yields

$$x = x^* \cos \varepsilon + y^* \sin \varepsilon = x^* \cos(-\varepsilon) - y^* \sin(-\varepsilon),$$
$$y = -x^* \sin \varepsilon + y^* \cos \varepsilon = x^* \sin(-\varepsilon) + y^* \cos(-\varepsilon),$$

and so property (ii) is satisfied.

(3) If

$$x^{**} = x^* \cos \delta - y \sin \delta, \qquad y^{**} = x^* \sin \delta + y^* \cos \delta,$$

then

$$
\begin{aligned}
x^{**} &= x^* \cos \delta - y^* \sin \delta, \\
&= (x \cos \varepsilon - y \sin \varepsilon) \cos \delta - (x \sin \varepsilon + y \cos \varepsilon) \sin \delta, \\
&= x(\cos \varepsilon \cos \delta - \sin \varepsilon \sin \delta) - y(\sin \varepsilon \cos \delta + \cos \varepsilon \sin \delta), \\
&= x \cos(\varepsilon + \delta) - y \sin(\varepsilon + \delta),
\end{aligned}
$$

as required. Similarly it can be shown that

$$y^{**} = x^* \sin \delta + y^* \cos \delta = x \sin(\varepsilon + \delta) + y \cos(\varepsilon + \delta),$$

and so property (iii) is satisfied.

(4) Property (iv) since addition in $\mathbb{C}$ is associative.

We remark that the reflection

$$x^* = -x, \qquad y^* = -y,$$

is a *discrete point transformation* that, although perhaps useful, does not constitute a one-parameter transformation group since it does not involve a continuous parameter with the requisite group properties.

Definition 2. A (continuous) *symmetry group of a differential equation* is a group of transformations that maps any solution to another solution of the equation.

Example 3. The homogeneous first-order ordinary differential equation

$$\frac{dy}{dx} = \frac{x^2 + y^2}{xy}, \tag{3}$$

has the solution

$$y(x) = x\sqrt{2(\ln x + c)}. \tag{4}$$

Making the transformation $x = x^* \, e^\alpha$, $y = y^* \, e^\alpha$, for which (3) is invariant, in (4) yields

$$y^*(x^*) = x^* \sqrt{2(\ln x^* + c + \alpha)},$$

which is another solution of (3). Thus, the transformation $x = x^* \, e^\alpha$, $y = y^* \, e^\alpha$ maps one solution of Equation (3) into another solution. This can also be obtained by making the transformation $y(x)/x = v(x)$.

Consider the one-parameter group of transformations $\mathcal{G}$

$$x^* = f(x, y; \varepsilon), \qquad y^* = g(x, y; \varepsilon), \tag{5}$$

with parameter $\varepsilon \in \mathbb{C}$. Expanding (5) about (the identity) $\varepsilon = 0$ yields

$$\begin{aligned} x^* &= x + \varepsilon\xi(x, y) + \mathcal{O}(\varepsilon^2), \\ y^* &= y + \varepsilon\phi(x, y) + \mathcal{O}(\varepsilon^2), \end{aligned} \tag{6}$$

where

$$\xi(x, y) = \left.\frac{df}{d\varepsilon}\right|_{\varepsilon=0}, \qquad \phi(x, y) = \left.\frac{dg}{d\varepsilon}\right|_{\varepsilon=0}, \tag{7}$$

since $f(x, y; 0) = x$ and $g(x, y; 0) = y$. Equation (5) is referred to as the *global* form of the group and (6) as the *infinitesimal* form. The crucial property of a one-parameter group of transformations is that given the infinitesimal form, one can deduce the global form by integrating

$$\frac{dx^*}{d\varepsilon} = \xi(x^*, y^*), \qquad \frac{dy^*}{d\varepsilon} = \phi(x^*, y^*), \tag{8}$$

subject to the initial conditions

$$x^* = x, \qquad y^* = y, \qquad \text{at } \varepsilon = 0. \tag{9}$$

This fundamental result that one can obtain the global form of the symmetry from the infinitesimal group is due to Lie. Essentially this is true since there is a unique local solution of first-order equations (8) subject to the initial conditions (9).

The vector (ξ, ϕ) is the *tangent vector*, at the point (x, y), to the curve described by the transformed points (x^*, y^*) and it is called the *tangent vector field* of the group $\mathcal{G}$.

Example 4. In this example we derive the infinitesimal form of the scaling group

$$x^* = x\,e^{\alpha\varepsilon}, \qquad y^* = y\,e^{\beta\varepsilon}, \tag{10}$$

and then by integration of (8) deduce the global form of the group. For the scaling group (10) we have

$$\frac{\mathrm{d}x^*}{\mathrm{d}\varepsilon} = \alpha x\,e^{\alpha\varepsilon}, \qquad \frac{\mathrm{d}y^*}{\mathrm{d}\varepsilon} = \beta y\,e^{\beta\varepsilon}$$

and so setting $\varepsilon = 0$ from (7) yields

$$\xi(x, y) = \alpha x, \qquad \phi(x, y) = \beta y.$$

Therefore, we need to integrate

$$\frac{\mathrm{d}x^*}{\mathrm{d}\varepsilon} = \alpha x^*, \qquad \frac{\mathrm{d}y^*}{\mathrm{d}\varepsilon} = \beta y^*,$$

subject to the initial conditions (9). Hence, it is seen that we recover the global form (10).

Now we turn our attention briefly to partial differential equations. In what follows let $(x, t) \in \mathbb{C}^2$ be the independent variables, $u \in \mathbb{C}$ the dependent variable. Then consider the scalar Nth order partial differential equation

$$\Delta = \Delta\left(x, t, u, u_x, u_t, u_{xx}, u_{xt}, u_{tt}, \ldots\right) = 0. \tag{11}$$

Suppose the set of solutions of Δ is given by $\mathcal{S} := \{u(x, t) : \Delta = 0\}$.

Definition 3. A *symmetry group* $\mathcal{G}$ of a system of differential equations Δ is a group of transformations that maps any solution to another solution of the system, i.e., the set $\mathcal{S}$ is mapped into itself.

Definition 4. A *Lie group of transformations* $\mathcal{G}$ is a group of transformation that depend continuously upon a parameter, or set of parameters.

Definition 5. A *Lie point transformation* is a transformation that depends only on the dependent and independent variables (and not on the derivatives of the dependent variables).

In this chapter we restrict our attention to local Lie point transformations, which have the form

$$x^* = X(x, t, u; \varepsilon), \qquad t^* = T(x, t, u; \varepsilon), \qquad u^* = U(x, t, u; \varepsilon), \qquad (12)$$

where $(x, t) \in \mathbb{C}^2$ and $u \in \mathbb{C}$ are the independent and dependent variables and ε is the group parameter. (Note that the functions X, T and U depend neither on the derivatives u_x, u_t nor higher-order derivatives.)

Before discussing the general theory for determining symmetry groups of differential equations, we consider some examples.

3. Ordinary differential equations

Example 5. Consider the homogeneous first-order ordinary differential equation

$$\frac{dy}{dx} = \frac{x^2 + y^2}{xy}, \qquad (13)$$

which we discussed in Example 3. This equation is usually solved by first making the substitution $y(x) = xv(x)$, which yields

$$x \frac{dv}{dx} + v = \frac{1 + v^2}{v},$$

and so

$$x \frac{dv}{dx} = \frac{1}{v}.$$

Then, by separating variables and integrating, one obtains the general solution of Equation (13)

$$y(x) = x\sqrt{2(\ln x + c)},$$

in which c is the constant of integration.

Why does the substitution $y(x) = xv(x)$ lead to a separable equation? Applying the transformation $x^* = x\,e^\alpha$, $y^* = y\,e^\beta$, with α and β real parameters, to Equation (13) yields

$$e^{\alpha - \beta} \frac{dy^*}{dx^*} = \frac{(x^*)^2\,e^{2\beta} + (y^*)^2\,e^{2\alpha}}{x^*y^*\,e^{\alpha + \beta}}. \qquad (14)$$

Therefore, if $\alpha = \beta$ then we see that Equations (13) and (14) are the same and so Equation (13) is said to be *invariant* under the transformation

$$x^* = x\,e^\alpha, \quad y^* = y\,e^\alpha, \qquad (15)$$

for *any* α. These transformations form a group, the so-called *group of scaling transformations* since the transformation (15) satisfies the properties to be a *one-parameter group of transformations*.

In this example, the substitution $v(x) = y(x)/x$ led to a separable equation since $v(x, y)$ is an *invariant* of the transformation (15), i.e.,

$$v(x^*, y^*) = \frac{y^*}{x^*} = \frac{y}{x} = v(x, y)$$

and it is this property which results in a simplification of Equation (3).

In general, if an ordinary differential equation is invariant under a one-parameter group of transformations, then the group leads to a simplification of the equation. If the equation is of first order then it becomes separable and so is solvable by quadrature, whilst for higher-order ordinary differential equations, then the use of an invariant permits a reduction in the order of the equation by one.

Example 6. The second-order equation for a simple pendulum is

$$\frac{\mathrm{d}^2 y}{\mathrm{d}x^2} + \sin y = 0, \tag{16}$$

which is invariant under the transformation

$$x^* = x + \varepsilon, \qquad y^* = y,$$

with ε a real parameter, for any ε. These mappings form a continuous group, the so-called *group of translations* along the x-axis, and thus setting

$$p = \frac{\mathrm{d}y}{\mathrm{d}x} = \frac{\mathrm{d}y^*}{\mathrm{d}x^*},$$

reduces Equation (16) to the first-order equation

$$p\frac{\mathrm{d}p}{\mathrm{d}y} + \sin y = 0.$$

This is integrated to give

$$\frac{1}{2}p^2 = \cos y + C,$$

with C an arbitrary constant, and so we obtain the first integral of Equation (16) given by

$$\frac{1}{2}\left(\frac{\mathrm{d}y}{\mathrm{d}x}\right)^2 = \cos y + C,$$

which can be further integrated in terms of elliptic functions using quadratures.

We remark that (16) is also invariant under the discrete group

$$y^* = y + 2n\pi, \qquad x^* = x, \qquad n \in \mathbb{Z}.$$

However, this is *not* a Lie group since n is a discrete parameter, not a continuous one. Further it is not clear how such discrete symmetries can be used to obtain solutions or reductions of differential equations.

Example 7. Consider the second-order equation

$$\frac{\mathrm{d}^2 y}{\mathrm{d}x^2} + \frac{\mu}{x}\frac{\mathrm{d}y}{\mathrm{d}x} - 2y^3 = 0, \tag{17}$$

with μ a constant. This is invariant under the transformation

$$x^* = x\,\mathrm{e}^{\alpha}, \qquad y^* = y\,\mathrm{e}^{-\alpha},$$

with α a real parameter. The invariant is

$$v = xy,$$

and so setting $y(x) = v(x)/x$ yields

$$x^2 \frac{\mathrm{d}^2 v}{\mathrm{d}x^2} + (\mu - 2)x\frac{\mathrm{d}v}{\mathrm{d}x} = (\mu - 2)v + v^3.$$

Now we let $v(x) = w(s)$, with $s = \ln x$, so that

$$\frac{\mathrm{d}^2 w}{\mathrm{d}s^2} + (\mu - 3)\frac{\mathrm{d}w}{\mathrm{d}s} = (\mu - 2)w + 2w^3,$$

which is an autonomous equation. This equation can then be reduced to a first-order equation, in an analogous way to the previous example.

In an analogous manner, symmetries can be used to reduce the number of equations in a system.

Example 8. Consider the system of equations

$$\frac{\mathrm{d}x}{\mathrm{d}t} = yz, \qquad \frac{\mathrm{d}y}{\mathrm{d}t} = zx, \qquad \frac{\mathrm{d}z}{\mathrm{d}t} = xy. \tag{18}$$

Since these equations are autonomous, they are equivalent to the system

$$\frac{\mathrm{d}y}{\mathrm{d}x} = \frac{x}{y}, \qquad \frac{\mathrm{d}z}{\mathrm{d}x} = \frac{x}{z}, \tag{19}$$

which have the respective solutions

$$y^2 = x^2 - C_1, \qquad z^2 = x^2 - C_2, \tag{20}$$

with C_1 and C_2 arbitrary constants. Substituting these into the first equation of Equation (18) yields

$$\left(\frac{dx}{dt}\right)^2 = (x^2 - C_1)(x^2 - C_2), \tag{21}$$

which is solvable in terms of Jacobi elliptic functions. Thus we have reduced a system of three first-order equations to one first-order equation.

4. Partial differential equations

The most common application of symmetry groups is to use them to reduce partial differential equations to ordinary differential equations. In principle, ordinary differential equations are easier to solve than partial differential equations.

Here, we are concerned with applications of symmetry groups to partial differential equations for which a one-parameter group of transformations permits a reduction in the total number of dependent and independent variables. We also give examples of some reductions that do *not* arise from a Lie point transformation. Typically a reduction of a scalar partial differential equation with two independent variables has the form

$$u(x,t) = F(x,t,w(z)), \qquad z = z(x,t),$$

where $w(z)$ satisfies an ordinary differential equation.

In this section, we shall primarily be interested in the following types of symmetry reductions (which are sometimes called similarity reductions) of partial differential equations

(a) *travelling waves*

$$u(x,t) = w(z), \quad z = x - ct,$$

with c a constant to be determined,

(b) *scaling reductions*

$$u(x,t) = t^\mu w(z), \quad z = xt^\lambda,$$

with μ and λ constants to be determined, and

(c) *accelerating waves*

$$u(x,t) = w(z) + \beta t^\mu, \quad z = x - \alpha t^2,$$

with α, β, and μ constants to be determined.

4.1.   *Travelling wave solutions*

In this subsection, we give several example of travelling wave reductions of partial differential equations. Travelling wave solutions occur with the partial differential equation is autonomous, i.e., it does not depend explicitly on the spatial variable x or the temporal variable t. Then the partial differential equation has both space translational and time-translational symmetries.

Example 9. The Korteweg–de Vries (KdV) equation

$$u_t + 6uu_x + u_{xxx} = 0, \tag{22}$$

is the prototypic example of a soliton equation solvable by the inverse scattering method [9]. Furthermore it has arisen in a number of physical contexts, including small amplitude surface gravity waves in shallow water, collision-free hydromagnetic waves, stratified internal waves, ion-acoustic waves, plasma physics and lattice dynamics, cf. Ablowitz and Clarkson [10], Ablowitz and Segur [11] and the references therein.

If we seek a travelling wave solution of the KdV equation (22) in the form

$$u(x, t) = w(z), \qquad z = x - ct,$$

then substituting this into (22) yields

$$w''' + 6ww' - cw' = 0.$$

Integrating this once gives

$$w'' + 3w^2 - cw = A,$$

with A an arbitrary constant, then multiplying by w'

$$w'w'' + 3w^2w' - cww' = Aw',$$

which on integration gives

$$\frac{1}{2}\left(w'\right)^2 + w^3 - \frac{1}{2}cw^2 = Aw + B, \tag{23}$$

with B another arbitrary constant. This equation is solved in terms of the Weierstrass elliptic function $\wp(z; g_2, g_3)$.

If we wish to obtain a solution $w(z)$ such that $w \to 0$ and $w' \to 0$ as $|z| \to \infty$, then necessarily $A = B = 0$ and so (23) becomes

$$(w')^2 = w^2(c - 2w), \qquad (24)$$

which yields the quadrature

$$\pm \int \frac{dw}{w\sqrt{c - 2w}} = z + z_0, \qquad (25)$$

with z_0 an arbitrary constant. Making the substitution $2w = c - v^2$, so $dw = -v\,dv$, in this yields

$$\frac{dw}{w\sqrt{c - 2w}} = 2 \int \frac{dv}{v^2 - c} = \frac{1}{\sqrt{c}} \ln \left(\frac{\sqrt{c} - v}{\sqrt{c} + v} \right),$$

then

$$v(z) = -\sqrt{c}\, \frac{\exp\{\sqrt{c}\,(z + z_0)\} - 1}{\exp\{\sqrt{c}\,(z + z_0)\} + 1} = -\tanh \left\{ \frac{1}{2}\sqrt{c}\,(z + z_0) \right\}$$

and so

$$w(z) = \frac{1}{2}\{c - v^2(z)\} = \frac{1}{2} \left\{ c - c\tanh^2 \left[\frac{1}{2}\sqrt{c}\,(z + z_0) \right] \right\}$$

$$= \frac{1}{2} c\,\text{sech}^2 \left\{ \frac{1}{2}\sqrt{c}\,(z + z_0) \right\}.$$

Therefore, we obtain the travelling wave reduction of the KdV equation (22)

$$u(x, t) = 2\kappa^2 \,\text{sech}^2\{\kappa(x - 4\kappa^2 t + \delta_0)\},$$

with $\kappa = \frac{1}{2}\sqrt{c}$ and $\delta_0 = \frac{1}{2}\sqrt{c}\,z_0$, which is the *soliton solution*.

Example 10. In this example, we discuss travelling wave reductions of the Boussinesq equation

$$u_{tt} + uu_{xx} + u_x^2 + u_{xxxx} = 0, \qquad (26)$$

which like the KdV equation (22) is a soliton equation solvable by inverse scattering [13, 14]. The Boussinesq equation arises in several physical applications: propagation of long waves in shallow water, one-dimensional nonlinear lattice-waves, vibrations in a nonlinear string and ion sound waves in a plasma.

If we seek a travelling wave solution of the Boussinesq equation (26) in the form

$$u(x,t) = w(z), \qquad z = x - ct,$$

then substituting this into (26) yields

$$w'''' + ww'' + (w')^2 + c^2 w'' = 0.$$

Integrating this twice gives

$$w'' + \frac{1}{2}w^2 + c^2 w = Az + B, \tag{27}$$

where A and B are arbitrary constants. This equation is solvable in terms of the first Painlevé equation

$$\frac{d^2 y}{dx^2} = 6y^2 + x, \tag{28}$$

(if $A \neq 0$) and elliptic or elementary functions (if $A = 0$).

4.2. Scaling reductions

Example 11. Suppose we seek a solution of the KdV equation (22) in the form

$$u(x,t) = t^\mu w(z), \qquad z = xt^\lambda,$$

where μ and λ are constants to be determined. Then

$$u_x = t^{\mu+\lambda} w', \qquad u_{xx} = t^{\mu+2\lambda} w'', \qquad u_{xxx} = t^{\mu+3\lambda} w''',$$
$$u_t = \mu t^{\mu-1} w + \lambda x t^{\mu+\lambda-1} w' = t^{\mu-1}(\mu w + \lambda z w')$$

and hence

$$u_t + 6uu_x + u_{xxx} = t^{\mu-1}(\mu w + \lambda z w') + 6t^{2\mu+\lambda} ww' + t^{\mu+3\lambda} w'''$$
$$= t^{\mu-1}\left(t^{3\lambda+1} w''' + 6t^{\mu+\lambda+1} ww' + \mu w + \lambda z w'\right) = 0.$$

This will give an ordinary differential equation for $w(z)$ provided that

$$3\lambda + 1 = 0, \qquad \mu + \lambda + 1 = 0 \quad \Rightarrow \quad \lambda = -\frac{1}{3}, \qquad \mu = -\frac{2}{3}.$$

Therefore, we obtain the scaling reduction of the KdV equation (22) given by

$$u(x,t) = t^{-2/3} w(z), \qquad z = xt^{-1/3}, \tag{29}$$

where $w(z)$ satisfies

$$w''' + 6ww' - \frac{2}{3}w - \frac{1}{3}zw' = 0. \tag{30}$$

This equation is solvable in terms of the second Painlevé equation

$$\frac{d^2y}{dx^2} = 2y^3 + xy + \alpha, \tag{31}$$

with α an arbitrary constant [12].

Example 12. Suppose we seek a solution of the Boussinesq equation (26) in the form

$$u(x,t) = t^\mu w(z), \qquad z = xt^\lambda,$$

where μ and λ are constants to be determined. Then

$$u_x = t^{\mu+\lambda}w', \qquad u_{xx} = t^{\mu+2\lambda}w'', \qquad u_{xxxx} = t^{\mu+4\lambda}w'''',$$
$$u_{tt} = t^{\mu-2}\left\{\mu(\mu-1)w + \lambda(\lambda+2\mu-1)zw' + \lambda^2 z^2 w''\right\},$$

and so

$$\begin{aligned}
u_{tt} &+ uu_{xx} + u_x^2 + u_{xxxx}\\
&= t^{\mu-2}\left\{\lambda^2 z^2 w'' + \lambda(\lambda+2\mu-1)zw' + \mu(\mu-1)w\right\}\\
&\quad + t^{2(\mu+\lambda)}\left[ww'' + (w')^2\right] + t^{\mu+4\lambda}w''''\\
&= t^{\mu-2}\left\{\lambda^2 z^2 w'' + \lambda(\lambda+2\mu-1)zw' + \mu(\mu-1)w\right.\\
&\quad \left. + t^{\mu+2\lambda+2}\left[ww'' + (w')^2\right] + t^{4\lambda+2}w''''\right\} = 0.
\end{aligned}$$

This is an ordinary differential equation for $w(z)$ provided that

$$\mu + 2\lambda + 2 = 0, \quad 4\lambda + 2 = 0 \quad \Rightarrow \quad \mu = -1, \quad \lambda = -\frac{1}{2}.$$

Therefore, we obtain the scaling reduction of the Boussinesq equation (26) given by

$$u(x,t) = t^{-1}w(z), \qquad z = xt^{-1/2}, \tag{32}$$

where $w(z)$ satisfies

$$w'''' + ww'' + (w')^2 + \frac{1}{4}z^2 w'' + \frac{7}{4}zw' + 2w = 0. \tag{33}$$

This equation is solvable in terms of the fourth Painlevé equation

$$\frac{d^2y}{dx^2} = \frac{1}{2y}\left(\frac{dy}{dx}\right)^2 + \frac{3}{2}y^3 + 4xy^2 + 2(x^2 - \alpha)y + \frac{\beta}{y}, \tag{34}$$

where α and β are arbitrary constants [15].

4.3.   *Accelerating wave solutions*

Example 13. Suppose we seek an accelerating wave solution of the KdV equation (22) in the form

$$u(x,t) = w(z) + \lambda t^m, \qquad z = x - \mu t^2,$$

where μ, λ and m are constants to be determined. Then

$$u_x = w', \qquad u_{xxx} = w''', \qquad u_t = -2\mu t w' + \lambda m t^{m-1}$$

and hence

$$
\begin{aligned}
u_t + 6uu_x + u_{xxx} &= -2\mu t w' + \lambda m t^{m-1} + 6(w + \lambda t^m)w' + w''' \\
&= (6\lambda t^m - 2\mu t)w' + \lambda m t^{m-1} + 6ww' + w'''.
\end{aligned}
$$

This will give an ordinary differential equation for $w(z)$ provided that $m = 1$ and $\mu = 3\lambda$. Hence, we obtain the accelerating wave reduction

$$u(x,t) = w(z) + \lambda t, \qquad z = x - 3\lambda t^2,$$

where $w(z)$ satisfies

$$w''' + 6ww' + \lambda = 0.$$

Integrating this once gives

$$w'' + 3w^3 + \lambda z = A,$$

with A an arbitrary constant, which is equivalent to the first Painlevé equation

$$\frac{d^2 y}{dx^2} = 6y^2 + xy, \qquad (35)$$

by rescaling the variables.

Example 14. Suppose we seek an accelerating wave solution of the Boussinesq equation (26) in the form

$$u(x,t) = w(z) + \lambda t^m, \qquad z = x - \mu t^2,$$

where μ, λ, and m are constants to be determined. Then

$$u_x = w', \qquad u_{xx} = w'', \qquad u_{xxxx} = w'''',$$
$$u_{tt} = 4\mu^2 t^2 w'' - 2\mu w' + \lambda m(m-1)t^{m-2}$$

and hence

$$
\begin{aligned}
u_{tt} + uu_{xx} &+ u_x^2 + u_{xxxx} \\
&= 4\mu^2 t^2 w'' - 2\mu w' + \lambda m(m-1)t^{m-2} + (w + \lambda t^m)w'' + (w')^2 + w'''' \\
&= w'''' + ww'' + (w')^2 + (4\mu^2 t^2 + \lambda t^m)w'' - 2\mu w' \\
&\quad + \lambda m(m-1)t^{m-2} = 0.
\end{aligned}
$$

This will give an ordinary differential equation for $w(z)$ provided that $m = 2$ and $\lambda = -4\mu^2$. Hence, we obtain the accelerating wave reduction

$$
u(x,t) = w(z) - 4\mu^2 t^2, \qquad z = x - \mu t^2, \tag{36}
$$

where μ is an arbitrary constant and $w(z)$ satisfies

$$
w''' + ww' - 2\mu w = 8\mu^2 z + A,
$$

with A an arbitrary constant, which is equivalent to (30) that arose in the description of scaling reductions of the KdV equation (22) and thus also solvable in terms of the second Painlevé equation (31).

5. Application of one-parameter transformation groups to partial differential equations

In this section, we discuss symmetry reductions of partial differential equations through one-parameter transformation groups using the KdV equation (22) and the Boussinesq equation (26) as illustrative examples.

5.1. *Korteweg–de Vries equation*

Case 1. Travelling wave solution

Consider the transformation

$$
x^* = x + x_0, \quad t^* = t + t_0,
$$

where x_0 and t_0 are arbitrary (nonzero) constants. Then it is seen that $x^* - ct^* = x - ct$, provided that $c = x_0/t_0$. Therefore, the KdV equation (22) is invariant under the one-parameter group of translations

$$
x^* = x + c\varepsilon, \quad t^* = t + \varepsilon, \quad u^* = u, \tag{37}
$$

where ε is the group parameter and c an arbitrary constant, for which the associated invariants are u and $x - ct$, i.e.,

$$x - ct = x^* - ct^* \quad \text{and} \quad u = u^*.$$

Setting these invariants to be the new independent and dependent variables, z and w respectively, yields the travelling wave reduction

$$u(x,t) = w(z), \qquad z = x - ct, \tag{38}$$

where $w(z)$ satisfies

$$w''' + 6ww' - cw' = 0.$$

Further, we note that if

$$u(x,t) = u^*(x^*, t^*), \qquad x^* = x + c\varepsilon, \qquad t^* = t + \varepsilon,$$

then

$$u_t + 6uu_x + u_{xxx} = u_{t^*}^* + 6u^*u_{x^*}^* + u_{x^*x^*x^*}^*.$$

Therefore, if $u(x,t)$ is *any* solution of the KdV equation (22), then so also is $u^*(x^*, t^*) = u(x,t)$, where $x^* = x + c\varepsilon$ and $t^* = t + \varepsilon$. The transformation (37) is a symmetry group, since it maps the set of solutions of the KdV equation into itself.

Case 2. Scaling reduction

If we apply the transformation

$$x^* = x\,e^{\alpha}, \qquad t^* = t\,e^{\beta}, \qquad u^* = u\,e^{\gamma},$$

to the KdV equation (22) then

$$e^{\beta - \gamma}u_t + 6\,e^{\alpha - 2\gamma}uu_x + e^{3\alpha - \gamma}u_{xxx} = 0,$$

i.e.,

$$u_t + 6\,e^{\alpha - \beta - \gamma}uu_x + e^{3\alpha - \beta}u_{xxx} = 0.$$

This is the same equation as (22) if and only if $\beta = 3\alpha$ and $\gamma = -2\alpha$. Therefore, setting $\alpha = \varepsilon$ we see that the KdV equation (22) is invariant under the one-parameter group of scaling transformations

$$x^* = x\,e^{\varepsilon}, \qquad t^* = t\,e^{3\varepsilon}, \qquad u^* = u\,e^{-2\varepsilon}, \tag{39}$$

where ε is the group parameter, for which the associated invariants are $x/t^{1/3}$ and $ut^{2/3}$, i.e.,

$$x/t^{1/3} = x^*/(t^*)^{1/3} \qquad \text{and} \qquad ut^{2/3} = u^*(t^*)^{2/3}.$$

These yield the scaling reduction

$$u(x,t) = t^{-2/3}w(z), \qquad z = x/t^{1/3}, \tag{40}$$

where $w(z)$ satisfies

$$w''' + 6ww' - \frac{2}{3}w - \frac{1}{3}zw' = 0. \tag{41}$$

Further, we note that if

$$u(x,t) = e^{2\varepsilon}u^*(x^*,t^*), \qquad x^* = x\,e^{\varepsilon}, \qquad t^* = t\,e^{3\varepsilon},$$

then

$$u_t + 6uu_x + u_{xxx} = e^{5\varepsilon}\left(u^*_{t^*} + 6u^*u^*_{x^*} + u^*_{x^*x^*x^*}\right).$$

Hence, if $u(x,t)$ is a solution of the KdV equation (22), then so is $u^*(x^*,t^*) = e^{-2\varepsilon}u(x,t)$, where $x^* = x\,e^{\varepsilon}$ and $t^* = t\,e^{3\varepsilon}$ and so the transformation (39) is a symmetry group, since it maps the set of solutions of the KdV equation into itself.

Case 3. Accelerating wave reduction

The KdV equation (22) is invariant under the one-parameter transformation group

$$x^* = x + 6\beta\varepsilon t + 3\beta\varepsilon^2, \qquad t^* = t + \varepsilon, \qquad u^* = u + \beta\varepsilon, \tag{42}$$

where ε is the group parameter and β an arbitrary (nonzero) constant; note that β does not appear in t^* and so is not a Lie group parameter. The associated invariants are $x - 3\beta t^2$ and $u - \beta t$, obtainable by eliminating ε in (42), i.e.,

$$x - 3\beta t^2 = x^* - 3\beta(t^*)^2 \qquad \text{and} \qquad u - \beta t = u^* - \beta t^*.$$

These yield the accelerating wave reduction

$$u(x,t) = w(z) + \beta t, \qquad z = x - 3\beta t^2,$$

where $w(z)$ satisfies

$$w'' + 3w^2 + \beta z = A,$$

where A is an arbitrary constant.

We remark that if

$$u(x,t) = u^*(x^*,t^*) - \beta\varepsilon, \qquad x^* = x + 6\beta\varepsilon t + 3\beta\varepsilon^2, \qquad t^* = t + \varepsilon,$$

then

$$u_t + 6uu_x + u_{xxx} = u_{t^*}^* + 6u^* u_{x^*}^* + u_{x^*x^*x^*}^*.$$

Hence, if $u(x,t)$ is a solution of the KdV equation (22), then so is $u^*(x^*,t^*) = u(x,t) + \beta\varepsilon$, where $x^* = x + 6\beta\varepsilon t + 3\beta\varepsilon^2$ and $t^* = t + \varepsilon$, and so the transformation (42) is a symmetry group, since it maps the set of solutions of the KdV equation into itself.

Further, we note that if $u(x,t)$ satisfies the KdV equation then so does $u^*(x^*,t^*) = u(x,t) + \beta$, where $x^* = x + 6\beta t$ and $t^* = t$, which is the so-called *Galilean transformation*.

5.2. *Boussinesq equation*

Case 1. Travelling wave solution

The Boussinesq equation (26) possesses the travelling wave solution

$$u(x,t) = w(z), \qquad z = x - ct, \tag{43}$$

where c is an arbitrary constant and $w(z)$ satisfies

$$w'' + c^2 w + \frac{1}{2}w^2 = Az + B, \tag{44}$$

with A and B arbitrary constants. An associated one-parameter group is

$$x^* = x + c\varepsilon, \qquad t^* = t + \varepsilon, \qquad u^* = u, \tag{45}$$

where ε is the group parameter, which has invariants $x - ct$ and u. If

$$u(x,t) = u^*(x^*,t^*), \qquad x^* = x + c\varepsilon, \quad t^* = t + \varepsilon,$$

then

$$u_{tt} + uu_{xx} + u_x^2 + u_{xxxx} = u_{t^*t^*}^* + u^* u_{x^*x^*}^* + (u_{x^*}^*)^2 + u_{x^*x^*x^*x^*}^*$$

and so the transformation (45) is a symmetry group since it maps the set of solutions of the Boussinesq equation into itself.

Case 2. Scaling reduction

The Boussinesq equation (26) possesses the scaling reduction

$$u(x,t) = t^{-1}w(z), \qquad z = x/t^{1/2}, \tag{46}$$

where $w(z)$ satisfies

$$w'''' + ww'' + (w')^2 + \frac{1}{4}z^2w'' + \frac{7}{4}zw' + 2w = 0.$$

An associated one-parameter group is

$$x^* = x\,\mathrm{e}^\varepsilon, \qquad t^* = t\,\mathrm{e}^{2\varepsilon}, \qquad u^* = u\,\mathrm{e}^{-2\varepsilon}, \tag{47}$$

where ε is the group parameter, which has invariants $x/t^{1/2}$ and ut. If

$$u(x,t) = u^*(x^*,t^*)\,\mathrm{e}^{2\varepsilon}, \qquad x^* = x\,\mathrm{e}^\varepsilon, \qquad t^* = t\,\mathrm{e}^{2\varepsilon},$$

then

$$u_{tt} + uu_{xx} + u_x^2 + u_{xxxx} = \mathrm{e}^{6\varepsilon}\left\{u^*_{t^*t^*} + u^*u^*_{x^*x^*} + (u^*_{x^*})^2 + u^*_{x^*x^*x^*x^*}\right\},$$

and so the transformation (47) is a symmetry group since it maps the set of solutions of the Boussinesq equation into itself.

Case 3. Accelerating wave reduction

It is known that the Boussinesq equation (26) also possesses the accelerating wave reduction [15–17]

$$u(x,t) = w(z) - 4\mu^2t^2, \qquad z = x - \mu t^2, \tag{48}$$

where μ is an arbitrary constant and $w(z)$ satisfies

$$w''' + ww' - 2\mu w = 8\mu^2 z + A,$$

with A an arbitrary constant. An associated one-parameter group is

$$x^* = x + 2\mu\varepsilon t + \mu\varepsilon^2, \qquad t^* = t + \varepsilon, \qquad u^* = u - 8\mu^2\varepsilon t - 4\mu^2\varepsilon^2, \tag{49}$$

where ε is the group parameter, which has invariants $x - \mu t^2$ and $u + 4\mu^2 t^2$. If

$$u(x,t) = u^*(x^*,t^*) + 8\mu^2\varepsilon t + 4\mu^2\varepsilon^2, \qquad x^* = x + 2\mu\varepsilon t + \mu\varepsilon^2, \qquad t^* = t + \varepsilon,$$

then

$$\begin{aligned}
u_{tt} + uu_{xx} &+ u_x^2 + u_{xxxx} \\
&= u^*_{t^*t^*} + u^*u^*_{x^*x^*} + (u^*_{x^*})^2 + u^*_{x^*x^*x^*x^*} + 4\mu\varepsilon u^*_{x^*t^*} + 8\mu^2\varepsilon t^* u^*_{x^*x^*},
\end{aligned}$$

which is *not* the Boussinesq equation (26) unless $\mu\varepsilon = 0$. Hence, the transformation (49) is *not* a (classical) symmetry group of the Boussinesq equation since it does not map the set of solutions into itself.

6. Discussion

The "classical Lie method" of symmetry reductions of a partial differential equation involves determining when the equation is invariant the Lie group (12). Rather than use the full form, one can use the infinitesimal form

$$x^* = x + \varepsilon\xi(x, t, u) + \mathcal{O}(\varepsilon^2), \tag{50a}$$

$$t^* = t + \varepsilon\tau(x, t, u) + \mathcal{O}(\varepsilon^2), \tag{50b}$$

$$u^* = t + \varepsilon\phi(x, t, u) + \mathcal{O}(\varepsilon^2). \tag{50c}$$

Requiring that the partial differential equation is invariant under the transformation (50) yields an overdetermined, linear system of equations for $\xi(x, t, u)$, $\tau(x, t, u)$ and $\phi(x, t, u)$; these are linear even if the partial differential equation under consideration is nonlinear. Having obtained a solution for ξ, τ and ϕ, one then solves the *invariant surface condition*

$$\xi(x, t, u)u_x + \tau(x, t, u)u_t = \phi(x, t, u), \tag{51}$$

to obtain the symmetry reduction; for further details and examples, see the books [2–7].

Motivated by the fact that the Boussinesq equation (26) also possesses the accelerating wave reduction (48), which does not arise as classical Lie symmetry, Clarkson and Kruskal [15] used a "direct method", to show that there are six canonical types of symmetry reductions for the Boussinesq equation (26):

$$u_1(x, t) = w_1(z), \qquad\qquad\qquad z = x + \mu_1 t,$$

$$u_2(x, t) = t^2 w_2(z) - x^2/t^2, \qquad\qquad z = xt,$$

$$u_3(x, t) = w_3(z) - 4\mu_3^2 t^2, \qquad\qquad z = x + \mu_3 t^2,$$

$$u_4(x, t) = t^2 w_4(z) - (x + 6\mu_4 t^5)^2/t^2, \qquad z = xt + \mu_4 t^6,$$

$$u_5(x, t) = t^{-1}w_5(z) - (x - 3\mu_5 t^2)^2/(4t^2), \qquad z = xt^{-1/2} + \mu_5 t^{3/2},$$

$$u_6(x, t) = \frac{1}{\wp(t)}\left\{w(z) - \left[\frac{1}{2}z\frac{d\wp}{dt} + \mu_6\wp^{3/2}(t)\right]^2\right\}, \qquad z = \frac{x + \mu_6\zeta(t)}{\wp^{1/2}(t)},$$

where μ_1, $\mu_3, \ldots, \mu_6$ are arbitrary constants, $\wp(t) = \wp(t + t_0; 0, g_3)$ is the Weierstrass elliptic function, $\zeta(t) = \zeta(t + t_0; 0, g_3)$ the Weierstrass zeta function, $w_1(z)$ and $w_2(z)$ satisfy an equation equivalent to the first Painlevé equation (35), $w_3(z)$ and $w_4(z)$ satisfy an equation equivalent to the second Painlevé equation (31) and $w_5(z)$ and $w_6(z)$ satisfy an equation equivalent to the fourth Painlevé equation (34). Only the travelling wave $u_1(x, t)$ and the scaling reduction $u_5(x, t)$ in the case when $\mu_5 = 0$ can be obtained by the classical Lie method. Levi and Winternitz [18] subsequently showed that these six symmetry reductions for the Boussinesq equation (26) can be obtained using the "nonclassical method" due to Bluman and Cole [19]; see also the review [20].

In the non-classical method it is required that the infinitesimal transformation (50) leaves invariant the set of simultaneous solutions of the partial differential equation and the invariant surface condition (51), where ξ, τ and ϕ are the same as in the transformation (50). In contrast to the classical method, these gives an overdetermined, nonlinear system of equations for the infinitesimals ξ, τ and ϕ. For some equations, such as the KdV equation (22), the symmetry reductions arising from the classical and non-classical methods are the same. During the past 25 years or so, the non-classical method has been successfully applied to obtain many new symmetry reductions and exact solutions for several physically significant partial differential equations.

References

[1] S. Lie, Zur allgemeinen Theorie derpartiellen Differentialgleichungen beliebeger Ordnung, *Leipz. Berich,* **1** (1895), 53–128.

[2] D. J. Arrigo, *Symmetry Analysis of Differential Equations: An Introduction,* Wiley, New York, 2015.

[3] G. W. Bluman and S. Kumei, *Symmetries and Differential Equations,* Springer-Verlag, Berlin, 1989.

[4] B. J. Cantwell, *Introduction to Symmetry Analysis,* C.U.P., Cambridge, 2002.

[5] J. M. Hill, *Differential Equations and Group Methods for Scientists and Engineers,* CRC Press, Boca Raton, 1992.

[6] P. E. Hydon, *Symmetry Methods for Differential Equations: A Beginner's Guide,* C.U.P., Cambridge, 2000.

[7] P. J. Olver, *Applications of Lie Groups to Differential Equations,* Second Edition, Springer-Verlag, New York, 1993.

[8] J. Carminati and K. Vu, Khai, Symbolic computation and differential equations: Lie symmetries, *J. Symbolic Comput.,* **29** (2000), 95–116.

[9] C. S. Gardner, J. M. Greene, M. D. Kruskal and R. M. Miura, Method for solving the KdV equation, *Phys. Rev. Lett.*, **19** (1967), 1095–1097.

[10] M. J. Ablowitz and P. A. Clarkson, *Solitons, Nonlinear Evolution Equations and Inverse Scattering*, C.U.P., Cambridge, 1991.

[11] M. J. Ablowitz and H. Segur, *Solitons and the Inverse Scattering Transform*, SIAM, Philadelphia, 1981.

[12] A. S. Fokas and M. J. Ablowitz, On a unified approach to transformations and elementary solutions of Painlevé equations, *J. Math. Phys.*, **23** (1983), 2033–2042.

[13] M. J. Ablowitz and R. Haberman, Resonantly coupled nonlinear evolution equations, *J. Math. Phys.*, **16** (1975), 2301–2305.

[14] V. E. Zakharov, On stocastization of one-dimensional chains of nonlinear oscillations, *Sov. Phys. JETP*, **38** (1974), 108–110.

[15] P. A. Clarkson and M. D. Kruskal, New similarity solutions of the Boussinesq equation, *J. Math. Phys.*, **30** (1989), 2201–2213.

[16] P. J. Olver and P. Rosenau, The construction of special solutions to partial differential equations, *Phys. Lett.*, **114A** (1986), 107–112.

[17] P. J. Olver and P. Rosenau, Group-invariant solutions of differential equations, *SIAM J. Appl. Math.*, **47** (1987), 263–275.

[18] D. Levi and P. Winternitz, Nonclassical symmetry reduction: example of the Boussinesq equation, *J. Phys. A*, **22** (1989), 2915–2924.

[19] G. W. Bluman and J. D. Cole, The general similarity of the heat equation, *J. Math. Mech.*, **18** (1969), 1025–1042.

[20] P. A. Clarkson, Nonclassical symmetry reductions of the Boussinesq equation, *Chaos, Solitons & Fractals*, **5** (1995), 2261–2301.

Printed in the United States
By Bookmasters